More Projects and Explorations

Writer: Michael Serra

Teacher's Materials Project Editor: Elizabeth DeCarli

Editor: Kendra Lockman

Project Administrator: Brady Golden

Contributors: Masha Albrecht, Dan Bennett, Ralph Bothe, Etsuo Hayashi, Judy Hicks, Darlene Pugh, Eberhard Scheiffele, Carolyn Sessions, Luis Shein, Sharon Grand Taylor

Reviewers: Christian Aviles-Scott, Larry Copes

Accuracy Checker: Dudley Brooks

Production Editor: Holly Rudelitsch

Copyeditor: Jill Pellarin

Editorial Production Manager: Christine Osborne

Production Supervisor: Ann Rothenbuhler

Production Coordinator: Jennifer Young

Text Designers: Jenny Somerville, Garry Harman

Composition, Technical Art, Prepress: ICC Macmillan Inc.

Cover Designers: Jill Kongabel, Marilyn Perry, Jensen Barnes

Printer: Von Hoffmann Corporation

Textbook Product Manager: James Ryan

Executive Editor: Casey FitzSimons

Publisher: Steven Rasmussen

Cover Photo Credits: Background image: Doug Wilson/Westlight/Corbis. Construction site image: Sonda Dawes/The Image Works. All other images: Ken Karp Photography.

Key Curriculum Press
1150 65th Street
Emeryville, CA 94608
(510) 595-7000
editorial@keypress.com
www.keypress.com

Printed in the United States of America
10 9 8 7 6 5 4 3 2 1 12 11 10 09 08 07 ISBN 978-1-55953-899-2

Contents

Below each project or exploration you will find a suggested *Discovering Geometry* lesson or chapter with which to use the activity. You may find another use that will better suit your needs. Refer to the Teacher's Notes that precede the activity for more detailed information.

Introduction

In this volume you will find activities—projects and explorations—collected from *Discovering Geometry* classrooms across the country. The activities are arranged by chapter. Some activities were written specifically for a particular chapter, while others are general-interest projects that you can use with any chapter of the book. Guided versions of some projects from the Student Edition of *Discovering Geometry* are also included. Each chapter begins with Teacher's Notes that suggest how to use the activities, and provide answers or possible outcomes.

You can register for Key Online at www.keypress.com/keyonline to get access to other *Discovering Geometry* Teacher Resources, including the following.

- *Logic Lessons:* These eight lessons from the Second Edition of *Discovering Geometry* provide a complete unit on symbolic logic in a lesson format, with examples and exercises. They cover the material on symbolic logic presented in the three logic Explorations in *Discovering Geometry,* Fourth Edition. Also included is a Logic Lessons Review, which can be used as an assessment of students' skills.
- *Cooperative Problem Solving activities:* This is a set of 16 group projects, also from the Second Edition of *Discovering Geometry*. Some are of general interest, but others extend material from a particular chapter. The activities share a common "outer space" theme. You can use them individually or as a set.

CHAPTER 0 TEACHER'S NOTES

PROJECT • Finding Geometry

Students find examples of geometry in nature and in art. This is a more guided version of the project in Lesson 0.6. It makes a good introduction to or review of Chapter 0. If your students are experienced computer users, you might encourage them to create Web pages or computer-based slideshows.

OUTCOMES

- Students present photographs or drawings with captions or give a computer based-presentation.
- The captions or the commentary mention geometry terms from this chapter.
- The use of terms is accurate.

PROJECT • Frank Lloyd Wright

Students use geometry tools to design a decorative panel or piece of furniture in the style of Frank Lloyd Wright, who used geometry extensively in his designs. You can use this project at any time during the course. If you use it with Chapter 8 or 10, you might ask students to calculate areas or volumes in their designs.

OUTCOMES

- Student has designed a decorative panel or piece of furniture.
- The report includes the geometric shapes and symmetries of the object and describes the tools used to create it.
- The report describes the geometry of *Fallingwater.*

Extra Credit

- The report includes research about other designs by Frank Lloyd Wright or other designers.
- Student designs the furnishings for a room.

PROJECT
Finding Geometry

Read over the four research topics below, then agree as a group how to divide the four tasks. Go to the library or look on the Internet to research your topic. Then meet as a group and have each group member present his or her findings to the group. Ask one another questions and suggest ways to improve one another's reports. Revise your work and hand it in as a group report or make a group presentation to the class.

1. The four-leaf clover is an example of a plant that exhibits the symmetry of a square. The pentagon (five-sided polygon) is represented in many flowers: spring violets, apple blossoms, wild roses, and forget-me-nots, for example. The hexagon (six-sided polygon) can be found in the petal arrangement of lilies, narcissus, jonquil, and asphodel. Make a photocopy or drawing of a flower with four, five, or six petals. Make sure you include the flower's name with the illustration. Write a paragraph describing geometric shapes and symmetries found in that plant.

2. Most nonmicroscopic animals that live on land (including humans) exhibit bilateral symmetry, but many sea creatures have more than bilateral symmetry. The starfish, for example, exhibits the rotational and reflectional symmetries of a regular pentagon. Make a photocopy or drawing of a sea animal that has more than bilateral symmetry. Write a paragraph or two describing its symmetries and naming any geometric shapes the animal resembles. Explain how having symmetries helps the animal survive.

3. Find a photograph of a design from a culture within the United States—perhaps an Amish quilt, a Hawaiian basket-weaving design, or a Native American blanket. Bring a photocopy or a drawing of the design to class. Write a paragraph describing what it is, what culture it's from, and what geometric shapes and symmetries are found in the design.

4. Find a photograph of a design from a culture outside the United States—perhaps a knot design from Africa, a mandala from Mexico, a Maori stitched tukutuku panel from New Zealand, or an Islamic wall tile design from Iran. Bring a photocopy or a drawing of the design to class. Write a paragraph describing what it is, what culture it's from, and what geometric shapes and symmetries are found in the design.

Now go out with your group and document examples of geometry in nature and art. Use a camera to take pictures of as many examples of geometry in nature and art as you can. Look for many different types of symmetry, and try to photograph art and crafts from many different cultures. Consider visiting museums and art galleries, but make sure it's okay to take pictures when you visit. You might find examples in your home or in the homes of friends and neighbors. If you take photographs, write captions for them that describe the geometric shapes and symmetries you find. If you take digital photos, you might create a Web page or a slideshow presentation.

PROJECT
Frank Lloyd Wright

Frank Lloyd Wright (1867–1959) is often called America's favorite architect. Some of his greatest contributions are in residential architecture. He built homes in 36 states: on mountaintops, nestled in woods, over streams, in the desert, in cities, and in suburbs.

Wright's architecture was based on elements of nature, and he called it organic architecture. *Fallingwater,* built over a waterfall near Mill Run, Pennsylvania (about 50 miles from Pittsburgh), demonstrates his philosophy of organic architecture. Look at the picture of *Fallingwater* on page 9 of your book. Wright said that *Fallingwater* was like a giant tree. What would represent the roots of the tree? The trunk of the tree? The branches of the tree? *Fallingwater* also displays an obvious love of geometry. Describe the geometry that you see in the picture.

Wright often designed not only the buildings, but also their skylights and art glass windows, their furniture, carpets, murals, and even their table settings.

Look in the library or on the Internet to find other designs by Frank Lloyd Wright. Then design an art glass panel, a wallpaper border pattern, or a piece of furniture in the style of Frank Lloyd Wright. Write a brief report about your design, including what geometric shapes and symmetries it has and the geometry tools you used to create it.

CHAPTER 1 TEACHER'S NOTES

GAME • Daffynition

Students practice writing definitions using random words from a dictionary. Students will see that dictionary definitions are *good definitions,* as defined in their book. You might use this game to supplement Lesson 1.3.

MATERIALS

- one dictionary per group

PROJECT • Non-Euclidean Geometry Portfolio

Students keep a year-long portfolio of their explorations into geometries on surfaces other than a Euclidean plane. Most of their work will probably be explorations on the surface of a sphere. If you have a heterogeneous class and are able to offer an honors option, this is an ideal extension for the honors credit. Or, you might offer extra credit for students who keep a portfolio of sphere explorations. Or, you might require work on the Lénárt Sphere as a project for the whole class. Whichever option you choose, students might build on Take Another Look activities in the student book and Extensions in the Teacher's Edition that mention explorations on spheres or other surfaces.

Your requirements for the portfolio might include students' reflections on their own thinking as they explore beyond the plane, and questions they themselves ask and seek answers to. Encourage students to ask their own questions and design their own explorations. A worksheet of sample questions that might lead to explorations for each chapter is included. Answers to these questions and many more possible explorations can be found in the book *Non-Euclidean Adventures on the Lénárt Sphere.*

MATERIALS

- sphere that can be drawn on, such as the Lénárt Sphere
- *Non-Euclidean Adventures on the Lénárt Sphere* by István Lénárt, *optional*
- Explorations worksheet, *optional*

EXPLORATION • Drawing a Four-Dimensional Hypercube

contributed by Dan Bennett

Being able to visualize or draw three-dimensional objects in two dimensions is an important skill in geometry. In this exploration students take that skill one step further to draw a four-dimensional hypercube. The fourth dimension is a topic that will certainly spark your students' interest and imagination. You might use this exploration at the end of Chapter 1, when students are practicing drawing three-dimensional objects.

For better or worse, we seem to live in a three-dimensional world—three dimensions of space are all we can see or move about in. Yet, theoretically, the universe could exist in more than three dimensions. Cosmologists, in fact, conjecture that the universe may exist in nine or ten spatial dimensions. These extra dimensions may just be "folded up" so that we can't perceive them. On the more practical side, higher-dimensional geometry is used to solve problems with many variables. Many complex problems in telecommunications, for example, have hundreds of variables. Computer programs that route telephone calls efficiently treat each variable as a dimension, as if simulating objects in hundreds of dimensions.

MATERIALS

- wire frame cube, *optional*

GUIDING THE ACTIVITY

You might demonstrate the different shadows of a cube by holding a wire frame cube over an overhead projector. However, the projector is not a parallel light source, so there will be some distortion.

QUESTIONS

1. 24 faces, 32 edges, and 16 vertices
2. Cubes; 8

EXTENSIONS

If you have Sketchpad, you might have students experiment with the sample sketch **Hypercube.gsp.** (Choose **File | Open | Samples | Sketches | Geometry | Higher Dimensions** or go to www.keypress.com/DG.) Students can rotate and animate the hypercube to give different views.

You might have students make models of the hypercube using straws held together with string or Zome system models.

REFERENCES

Flatland by Edwin A. Abbott

The Mathematical Tourist by Ivars Peterson

Zome Geometry by George Hart and Henri Picciotto (Unit 21)

Exploring the Shape of Space by Jeffrey R. Weeks

For complete references to these and other sources, see www.keypress.com/DG.

GAME • Geopardy

contributed by Judy Hicks

This game is adapted from the popular television game show. It is an alternative way to review for a chapter test or a semester final. The material in this book is a complete set of materials for a Chapter 1 Geopardy game. You can use this as a model to write a Geopardy game as a chapter review for any other chapter in the book. You might also have groups of students make up question-and-answer cards for particular subjects or particular chapters, and you may want to add a few trivia questions that pertain only to your class. When your students make up their questions, they should also write the answer to each question on the same card. This way you will have many different cards to choose from. However, take care that their questions are not too difficult or confusing.

Depending on your computer experience, this entire activity can also be done on the computer. Presentation software can be used to simulate the game. All scoring and so on can be done electronically. Music and other sound effects can easily be included for fun. The computer can then be connected to a projector for all the students to see. Another variation is to have several computers set up and have students play against each other at each computer.

Materials

- gameboard transparency with the categories and points
- set of review questions and answers
- one blank Geopardy scorecard for each student

How to Play

1. Divide the class into three teams.
2. Give each student a blank Geopardy scorecard (big enough for work to be shown).
3. Assign each student a random number between 1 and the number of students in the class. You can easily do this with a calculator ahead of time.
4. Roll a die (or use the calculator) to determine which team starts. Use the calculator again to randomly choose a student on that team to pick a category and value from the gameboard.
5. Read a review question.
6. Make sure all the students work each problem.
7. The student chosen may confer with his or her team before answering. If the answer is correct, that team receives points. If the answer is incorrect, another team has the chance to answer. You may choose to discuss some questions now or later.
8. If the team answers the question correctly and receives points, the team score is adjusted accordingly. Post each team's overall score on the board.
9. For the next question, again randomly pick a new team and person, or just go around in order and follow the same directions.
10. The gameboard should include two "daily doubles," announced by you when a student makes that selection. For a "daily double," the team and student chosen may bet a maximum of 50 points if their score is less than the amount of the selection picked. If their score is more than 50, they can only bet up to double the amount of the selection. If their answer is correct, the overall score is increased by double the bet. If it is incorrect, their score is decreased by the amount of the bet.
11. When there are about 10 minutes left, every team and member plays "Final Geopardy." Each team is notified of the category and asked to write the amount of its bet on its scorecard. Each team must cooperatively make this decision. The team may only bet up to the amount of points it has in its score (or 50 points, if it has less than 50). The question is then displayed. After a certain amount of time, each team must decide on the answer, and its members must write the answer on their scorecards. All scorecards are then passed in. Choose one scorecard from each team and read the answer, then add or subtract the points from the team's score accordingly.
12. The winning team's members receive a "bonus buck," which they can use to excuse tardiness, as bonus points or assignment points, or for candy, and so on.

GAME
Daffynition

This game will sharpen your definition-writing skills. The object is to write fake, but convincing, definitions. As you play, you'll discover that a dictionary definition has the characteristics of good definitions as you've learned them: It's succinct, it places the defined object in a class, and it describes what differentiates that object from other objects in its class. Your fake definitions should have those characteristics, too. Now, on with the game!

How to Play

1. Divide into groups of four or five.
2. Select a scorekeeper for your group. Then choose a person to start the game. Call this person the selector.
3. To begin a round, the selector finds a strange new word in the dictionary. It must be a word that nobody in the group knows. (If you know the word, you should say so, and the selector will pick a new word.)
4. Each person then writes down a fake definition that sounds real. The selector copies the real definition from the dictionary.
5. The selector collects all the fake definitions and mixes them up with the real definition. The selector numbers the definitions.
6. The selector then reads all the definitions out loud, trying to make each one sound real.
7. Everyone except the selector writes on a "guess" card the number of the definition she or he thinks is the real one. You cannot choose your own definition. All players show their guesses at the same time.
8. Score each round as follows: If you're the selector, you get one point for each person who guessed the wrong definition. If you're not the selector, you get one point if you guessed correctly. You get two points for each person who guessed your fake definition.
9. After scoring, the person to the selector's right becomes the selector for the next round. The game continues until everyone has had a chance to be the selector.

Sample Scorecard

	Round 1	Round 2	Round 3	Round 4	Total
Player A					
Player B					
Player C					
Player D					

PROJECT

Non-Euclidean Geometry Portfolio

In this course you'll study many geometrical figures that lie in a flat plane, like a piece of paper. This is called Euclidean geometry. However, you can also study what geometry is like on the surface of a sphere, like the Earth, or on other shapes, like a cone or cylinder. Sometimes the properties that are true for a figure in the plane will be true on other surfaces and sometimes they'll be completely different.

For this project, you'll keep a reflective portfolio of your explorations of spherical and other non-Euclidean geometries. You and your teacher will discuss the scope and possible content of your project. You will find some good questions to start your thinking in the Take Another Look sections in some Chapter Reviews. Use these and other things your teacher suggests as you explore and compare geometry on a plane and geometry on a sphere.

A portfolio is a record of work that shows how your ideas grew and changed as you learned more. A reflective portfolio also includes your thoughts about what you learned.

Each entry in your portfolio should include:

1. **A problem or question** you explored on a sphere or other surface. Use a sphere you can draw on, such as the Lénárt Sphere.
2. **The process you followed** as you answered the question. You will want to include some of the things you tried and thought would work that turned out to be wrong. You may have learned the most from these.
3. **Your solutions,** conjectures, or conclusions. Make sure your solution comes from the process you described and includes an answer to the question "Why?"
4. **Further questions** your exploration caused you to think about. You might include your work on these questions or come back to these same questions later in your portfolio.
5. **Your thinking** about your process, your conclusions, and your questioning. You might relate your thinking about non-Euclidean geometries to what you are learning about geometry on a plane.

You and your teacher can discuss ways you can indicate each of these five parts throughout your portfolio.

PROJECT

Non-Euclidean Geometry Portfolio

Explorations

There are many questions that you can investigate on spheres and other surfaces. Here are a few to get you started.

Chapter 0 Extend to the surface of a sphere one geometric-art idea, such as line designs or tiling.

Chapter 1 Write definitions of lines and other geometric figures on a sphere.

How do those definitions compare to the definitions for plane geometry?

Chapter 2 What do perpendiculars look like on a sphere?

Chapter 3 What are lines and parallel lines on the surface of an infinite cylinder?

How many perpendiculars can two lines have in common on a cylinder?

Chapter 4 What triangle conjectures hold for triangles on a sphere? (See Take Another Look activities 1 and 2 on page 255 of your book.)

Chapter 5 How could you define a regular "sphere polygon"?

Which regular sphere polygons are possible and why? (See Take Another Look activity 2 on page 307 of your book.)

Is it possible to make a polygon with only two sides on a sphere?

Chapter 6 What conjectures appear to be true for circles on a sphere?

Chapter 7 How would you create a spherical tessellation with reflective symmetry?

Chapter 8 Are there formulas for the surface areas of "sphere polygons"? Explain.

Chapter 9 How could you extend the Pythagorean Theorem to triangles on a sphere? Explain.

Chapter 10 What would geometry on the surface of an infinite cone look like? Explain.

Chapter 11 What can you discover about similar triangles on a sphere? Explain.

Chapter 12 What is spherical trigonometry?

Chapter 13 Write a system of postulates and theorems for spherical geometry. (See the Exploration Non-Euclidean Geometries on pages 742–744 of your book.)

EXPLORATION

Drawing a Four-Dimensional Hypercube

You're used to drawing three-dimensional objects in two dimensions. In this activity you'll take that skill one step—or dimension—further: You'll see how to draw a *four*-dimensional object called a *hypercube.*

Activity: The Hypercube

Before we go off into the fourth dimension, let's look at the steps you might go through to draw a familiar, three-dimensional cube.

Steps 1–3 Start with a zero-dimensional object, a point. Now go in any direction you choose for the first dimension, copy your point, and connect the two points to create a one-dimensional object, a line segment. The arrow represents the direction of the first dimension.

Step 1 Step 2 Step 3

Steps 4–6 Now go the same distance in the direction of the second dimension, perpendicular to the direction you chose for the first dimension, and copy your segment. Connect the endpoints and you have a two-dimensional object, a square.

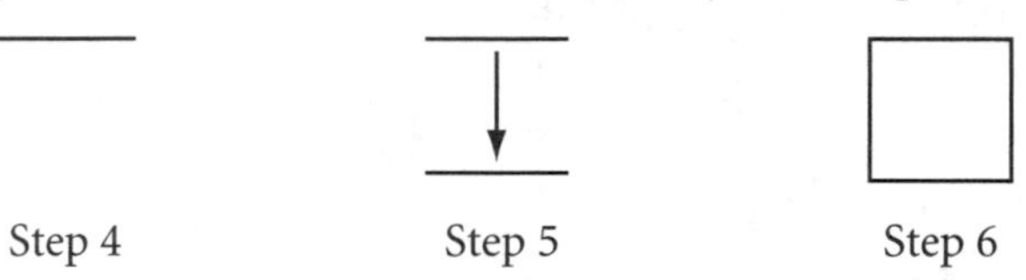

Now things start to get interesting. The direction of the third dimension, if it were to be perpendicular to your first two dimensions, would have to come directly out of the page. But when you draw three-dimensional objects on a two-dimensional page, you choose the direction for the third dimension arbitrarily.

Steps 7–9 Choose a direction for the third dimension, move that direction, and copy your square. Connect corresponding vertices of your two squares and you have a three-dimensional object, a cube.

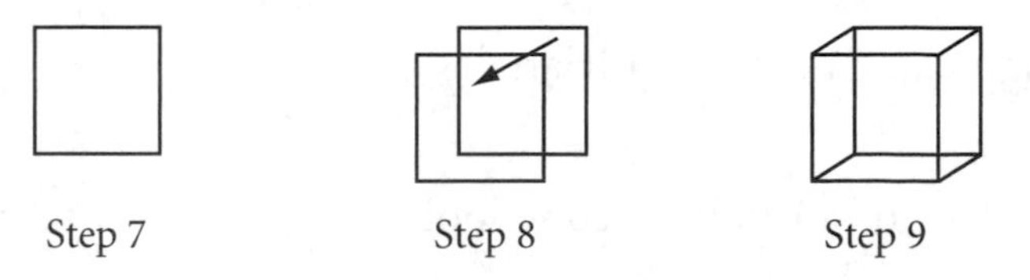

What you actually have is a two-dimensional representation of a three-dimensional object. An actual cube has perpendicular edges, and many of the segments in the above drawing are not perpendicular. But a viewer can figure out what you mean and see your "cube."

(continued)

Exploration • Drawing a Four-Dimensional Hypercube (continued)

One way to think of a two-dimensional representation of a cube is to imagine it's the shadow cast by a wire frame. If you take a wire frame cube outside and cast its shadow on a surface perpendicular to the Sun's rays, you can create a shadow just like the figure above. And if you change the orientation of your cube, you can create other shadows that might be surprising.

All the above figures are shadows of cubes. They're two-dimensional representations of the same three-dimensional object. The only difference is the direction chosen for the third dimension. Some of the representations you probably wouldn't recognize as cubes, but that's just because you're not used to seeing cubes drawn that way.

Step 10 Now choose a direction for the fourth dimension. Move that direction and copy your cube. (Try tracing it.) Connect corresponding vertices of your two cubes. You have a two-dimensional representation of a four-dimensional object, a hypercube.

Below are some different views of hypercubes. Can you tell which was the original cube? Which direction was chosen for the fourth dimension? The last example shows one way hypercubes are often drawn, as a cube within a cube. In this case, the fourth dimension was chosen to move in every direction away from the center of the cube.

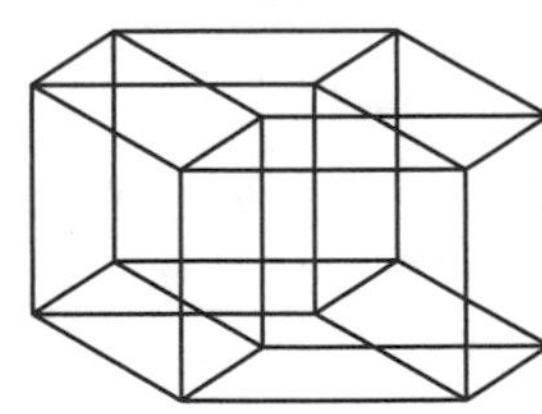 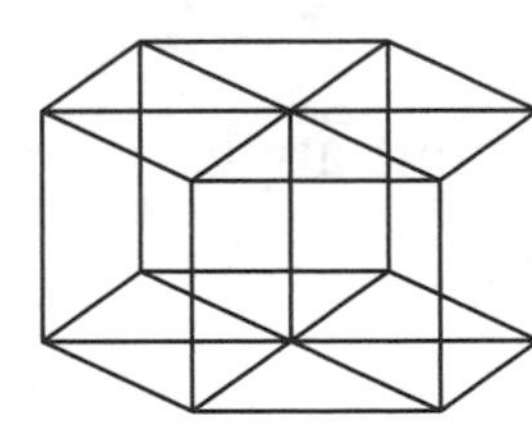 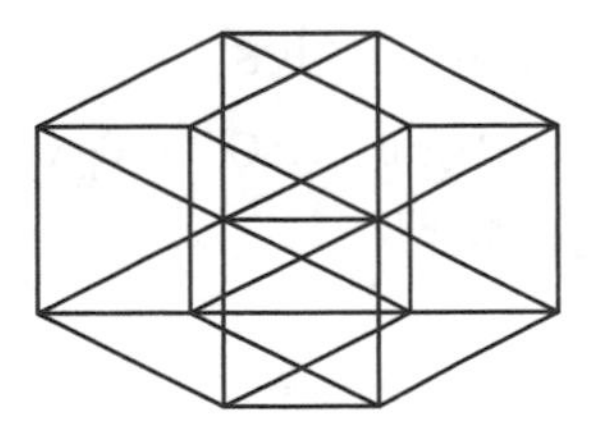 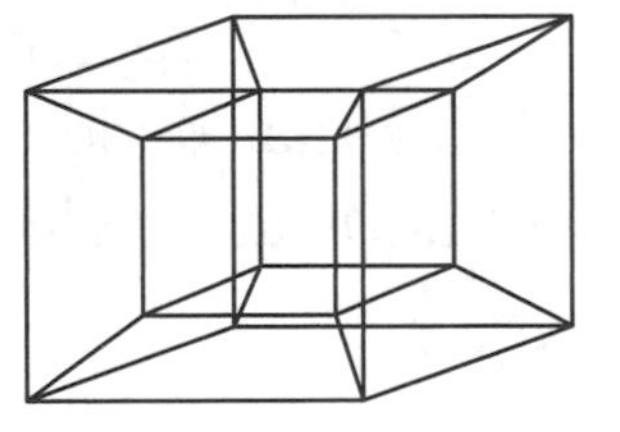

Questions

1. How many faces, edges, and vertices does a hypercube have? Rather than simply counting, think about the process you used to draw a hypercube.

2. Vertices are zero-dimensional points, edges are one-dimensional segments, and faces are two-dimensional squares. What three-dimensional objects are parts of a hypercube? How many of these parts does a hypercube have?

GAME

Geopardy (page 1 of 5)

Sample Questions for Chapter 1

Angles for

10 An angle whose measure is 90 degrees
What is a right angle?

20 An angle whose measure is less than 90 degrees
What is an acute angle?

30 An angle whose measure is greater than 90 degrees
What is an obtuse angle?

40 Two angles with the same measure
What are congruent angles?

50 An example that proves a statement wrong
What is a counterexample?

Angles and Lines for

10 Two or more lines that lie in the same plane and do not intersect
What are parallel lines?

20 Two lines that intersect to form a right angle
What are perpendicular lines?

30 Two angles whose measures have the sum of 90 degrees
What are complementary angles?

40 Two angles whose measures have the sum of 180 degrees
What are supplementary angles?

50 A ray that divides an angle into two congruent angles
What is an angle bisector?

Polygons for

10 A five-sided polygon
What is a pentagon?

20 A polygon whose sides are equal in measure
What is an equilateral polygon?

30 A polygon in which no diagonal is outside the polygon
What is a convex polygon?

40 A ten-sided polygon
What is a decagon?

50 Two sides of a polygon that share a common vertex
What are consecutive sides?

Sample Questions for Chapter 1

Triangles for

10 A triangle with one right angle
What is a right triangle?

20 A triangle with three acute angles
What is an acute triangle?

30 A triangle with one obtuse angle
What is an obtuse triangle?

40 A triangle with at least two congruent sides
What is an isosceles triangle?

50 Two angles of a triangle that are opposite two sides of equal length
What are base angles?

Quadrilaterals for

10 A quadrilateral with two pairs of opposite parallel sides
What is a parallelogram?

20 An equilateral parallelogram
What is a rhombus?

30 A quadrilateral with two distinct pairs of consecutive congruent sides
What is a kite?

40 A quadrilateral with exactly one pair of parallel sides
What is a trapezoid?

50 An equiangular rhombus or an equilateral rectangle
What is a square?

Circles for

10 A segment from the center of a circle to a point on the edge of the circle
What is a radius?

20 Two points on a circle and the continuous part of the circle between them
What is an arc of a circle?

30 A line segment whose endpoints lie on the circle
What is a chord?

40 Two or more coplanar circles with the same center
What are concentric circles?

50 A line that intersects a circle only once
What is a tangent?

Geopardy (page 3 of 5)

Sample Questions for Chapter 1

Potpourri for

10 Two or more points that lie on the same line
What are collinear points?

20 A unit of measure for angles
What is a degree?

30 Lines that are not in the same plane and that do not intersect
What are skew lines?

40 A closed geometric figure in a plane in which line segments connect endpoint to endpoint and each segment intersects exactly two others
What is a polygon?

50 An 11-sided polygon
What is an undecagon?

Final Geopardy

A polygon with at least one diagonal outside the polygon
What is a concave polygon?

Sample Scorecard for Chapter 1

Geopardy

Angles	Angles and Lines	Polygons	Triangles	Quads	Circles	Potpourri

Name ________________ Final Geopardy Answer ________________

Team ________________ Bet ________________

Geopardy (page 5 of 5)

Sample Gameboard for Chapter 1

Geopardy

Angles	Angles and Lines	Polygons	Triangles	Quads	Circles	Potpourri
10	10	10	10	10	10	10
20	20	20	20	20	20	20
30	30	30	30	30	30	30
40	40	40	40	40	40	40
50	50	50	50	50	50	50

PROJECT • Polygonal Numbers

Students extend their reasoning about triangular and rectangular numbers in Lesson 2.3 to other figurate numbers.

OUTCOMES

- Student gives correct rule for pentagonal numbers: $\frac{n(3n-1)}{2}$
- Correct rules and logical reasoning are shown for at least three other sets of polygonal numbers. Sets might include

  hexagonal numbers: $n(2n - 1)$

  heptagonal numbers: $\frac{n(5n - 3)}{2}$

  octagonal numbers: $n(3n - 2)$

  nonagonal numbers: $\frac{n(7n - 5)}{2}$

  decagonal numbers: $n(4n - 3)$
- Sample observations: The odd-numbered rules are similar to the even-numbered rules, except they are divided by 2. The coeffecients in the even-numbered rules increase by 1 as the number of sides in the polygon increases, and the coeffecients in the odd-numbered rules increase by 2.

PROJECT • Beehive Geometry

Students look for patterns in sequences of connected hexagons. This project extends the patterns students find as they observe sequences of squares and rectangles in Lessons 2.1 and 2.2.

MATERIALS

- Hexagonal Grid worksheet

OUTCOMES

1. 322 non-queen cells *(See table at bottom of page.)*
2. 28 queen cells, 60 non-queen cells *(See table at bottom of page.)*
3. 28 queen cells, 114 non-queen cells *(See table at bottom of page.)*
4. 60 queen cells on the boundary, 19 queen cells on the diagonal, 79 queen cells altogether *(See table at bottom of page.)*
5. Student designs a beehive pattern and gives a correct rule for it.

Project · Beehive Geometry, Outcomes 1, 2, 3, 4

1.

Number of queen cells	1	2	3	4	5	6	7	n
Number of non-queen cells	6	10	14	18	22	26	30	$4n + 2$

2.

Week	1	2	3	4	5	6	7	n
Number of queen cells	1	4	7	10	13	16	19	$3n - 2$
Number of non-queen cells	6	12	18	24	30	36	42	$6n$

3.

Week	1	2	3	4	5	6	7	n
Number of queen cells	1	4	7	10	13	16	19	$3n - 2$
Number of non-queen cells	6	18	30	42	54	66	78	$12n - 6$

4.

Week	1	2	3	4	5	6	7	n
Number of queen cells	7	15	23	31	39	47	55	$8n - 1$
Number of non-queen cells	0	4	14	30	52	80	114	$3n^2 - 5n + 2$

PROJECT • Euler's Formula for Networks

Students discover Euler's formula for the number of points, edges, and regions in a network. This project extends the Exploration The Seven Bridges of Königsberg. You might also use it with the Chapter 10 Exploration Euler's Formula for Polyhedrons. The two formulas are equivalent.

OUTCOME

- Student gives the correct rule: $P + R = E + 2$, where P is the number of points, R is the number of regions, and E is the number of edges, or paths.

EXPLORATION • Map Coloring

contributed by Masha Albrecht and Darlene Pugh

Students discover the Four-Color Theorem. If you haven't done so yet, you may want to define **theorem** (a conjecture that has been proved). This term will come up in Lesson 9.1 on the Pythagorean Theorem. This exploration consists of five parts and is really a collection of five short projects that build in difficulty. You can choose one project from this collection or do them all in the order in which they appear. This exploration can be used any time during the course.

If students have done the Exploration The Seven Bridges of Königsberg, they may connect the rules for chromatic numbers with the rules for traveling a network.

MATERIALS

- crayons or pencils of at least six different colors (Parts 1–3)
- Map of the United States worksheet (Part 1)
- 1-meter piece of string (Part 2)
- cellophane tape (Part 2)
- unlined paper (Parts 2 and 3)
- research materials from a library or the Internet (Part 4)
- torus (Part 5). If your classroom has a Lénárt Sphere kit, the kit contains a torus. You can draw maps on this torus using dry-erase pens.

OUTCOMES

Part 1

1. 4

2. 4

Part 2

1. 2

2. 2

3. Sample answer: Any number of intersecting lines (without endpoints) will divide the paper into an even number of regions, with no odd points. These regions can then be colored in a two-color checkerboard pattern. Similarly, the string doesn't have any endpoints, so it always divides the paper into an even number of regions with no odd points.

Part 3

1. Sample map:

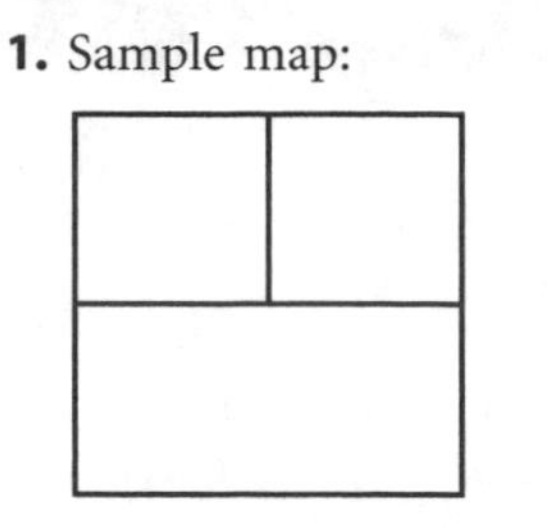

2. Sample map:

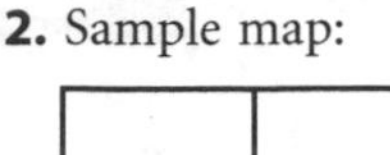

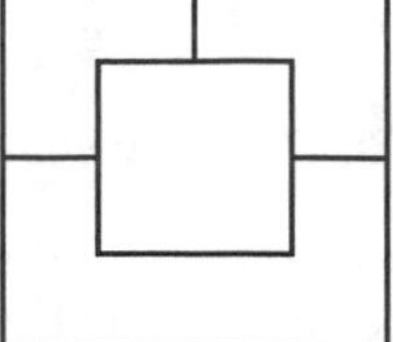

3. This is impossible in the plane.

Part 4

- Report summarizes the history of the Four-Color Theorem.

Extra Credit

- Report includes recent research about special cases in which the theorem may not apply. For instance, this map of six countries, two with infinitely many zigzags as they approach the center boundary, cannot be colored with four colors.

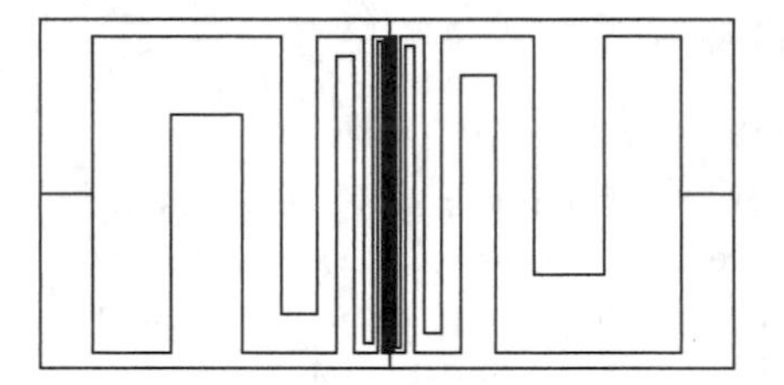

Part 5

- The largest chromatic number on a torus is seven.

PROJECT

Polygonal Numbers

Ancient mathematicians, most notably the Alexandrian mathematician Diophantus (ca. 210–ca. 290 C.E.), were particularly interested in numbers that corresponded to geometric figures. Such numbers are called **figurate numbers.** The most recognizable of these are the square numbers: 1, 4, 9, 16, and so on. In Lesson 2.3, you also learned about triangular and rectangular numbers, and found function rules to represent them. What about other polygons?

Here are the first four pentagonal numbers. Make a table and find a rule that generates the numbers. (*Hint:* Try factoring them. Keep in mind how you found the rule for the triangular numbers.)

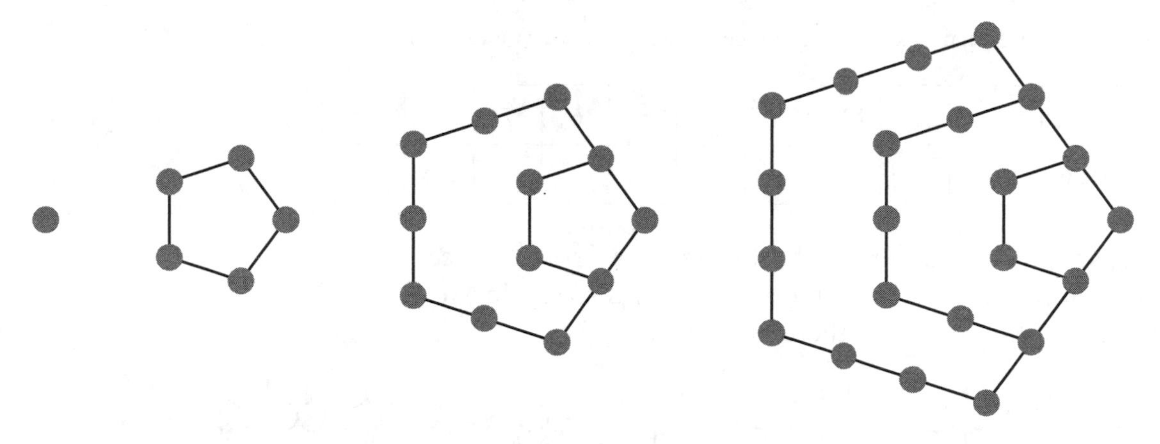

Now draw at least three other sets of polygonal numbers and find rules for them. Explain how you found each rule. Note any patterns or similarities you see among the rules.

PROJECT

Beehive Geometry

In Chapter 2, you looked for patterns in sequences of squares and rectangles. In this project you will look for patterns in sequences of hexagons. Pretend that the darker-colored hexagons are potential homes for queen bees in a honeycomb.

1. If the pattern of hexagons continues in the hexagon grid, how many non-queen cells will there be when there are 80 queen cells?

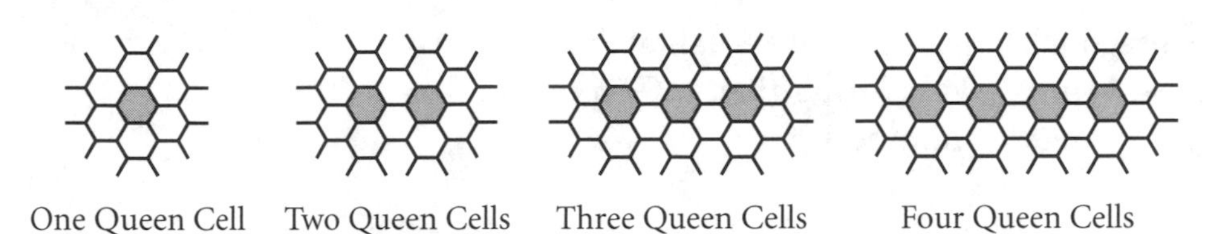

One Queen Cell Two Queen Cells Three Queen Cells Four Queen Cells

Number of queen cells	1	2	3	4	5	6	7	. . .	n
Number of non-queen cells								. . .	

2. This pattern shows the growth of a beehive for the first four weeks. If the pattern continues, how many queen cells will there be in Week 10? How many non-queen cells will there be in Week 10?

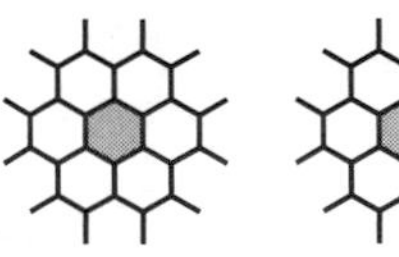

Week 1

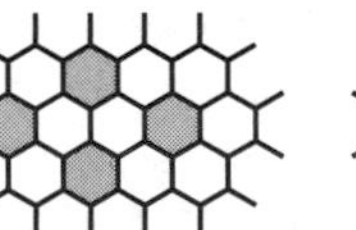

Week 2

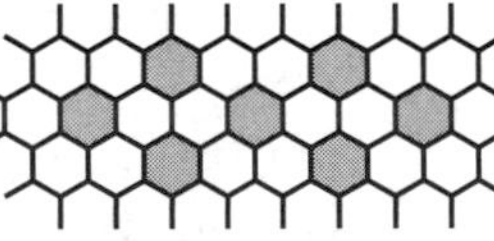

Week 3

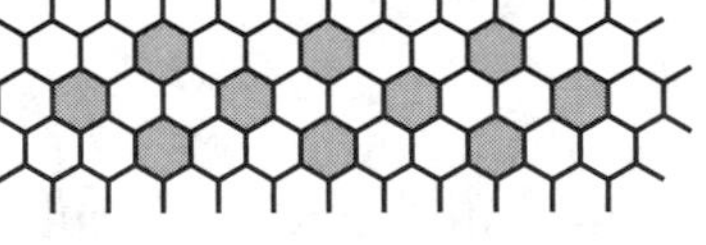

Week 4

Week	1	2	3	4	5	6	7	. . .	n
Number of queen cells								. . .	
Number of non-queen cells								. . .	

(continued)

3. This pattern shows the growth of a beehive for the first four weeks. The worker bees working on this honeycomb are not as linear in their thinking as those in Steps 1 and 2. They are building their pattern of queen cells in three directions. If the pattern continues, how many queen cells will there be in Week 10? How many non-queen cells in Week 10?

Week 1 Week 2 Week 3 Week 4

Week	1	2	3	4	5	6	7	...	n
Number of queen cells								...	
Number of non-queen cells								...	

4. This pattern shows the growth of a beehive for the first four weeks. The worker bees working on this honeycomb are building near a "Do Not Enter" sign on a one-way street. The worker bees decided to build their honeycomb pattern to tell humans, "Do not enter here!" If the pattern continues, how many queen cells will there be on the boundary, or perimeter, in Week 10? How many queen cells will there be on the diagonal in Week 10? How many queen cells will there be altogether in Week 10?

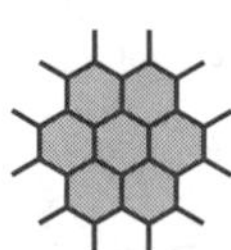

Week 1

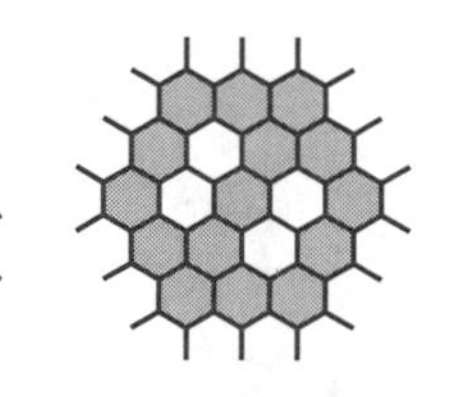

Week 2

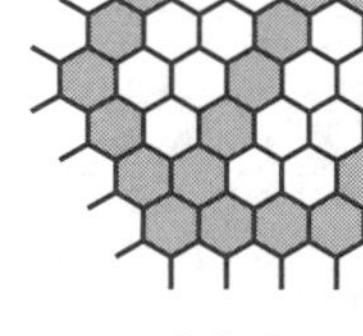

Week 3

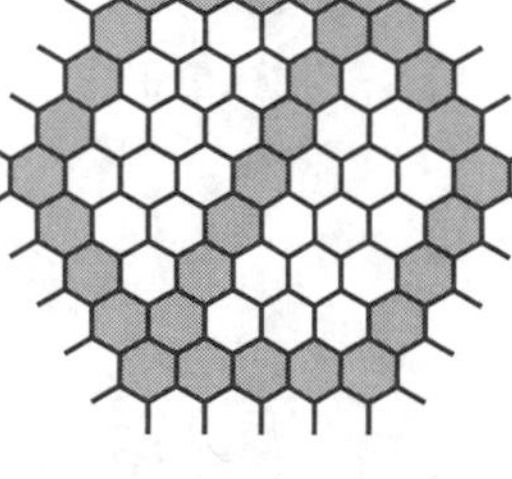

Week 4

Week	1	2	3	4	5	6	7	...	n
Number of queen cells								...	
Number of non-queen cells								...	

5. On hexagonal grid paper, design the first four terms of your own beehive pattern. Find a rule for your pattern.

PROJECT
Beehive Geometry

Hexagonal Grid

PROJECT
Euler's Formula for Networks

In the Exploration The Seven Bridges of Königsberg, you saw that networks consist of points and paths, or edges. Connecting points with edges forms regions between the edges. There is a formula for the relationship among the numbers of points, edges, and regions in a network, and this formula holds for any network that can be drawn on a piece of paper (not just the networks that can be traveled). Your job is to discover that formula. To help you start, make a table counting the numbers of points, edges, and regions for each network below. When counting regions, include the region that remains on the outside of the network. (The reason for counting the outer region will become clear when you find Euler's Formula for Polyhedrons in Chapter 10.) Using your table, look for a relationship among the three parts of a network. Once you've found a formula, test it by drawing a few networks of your own.

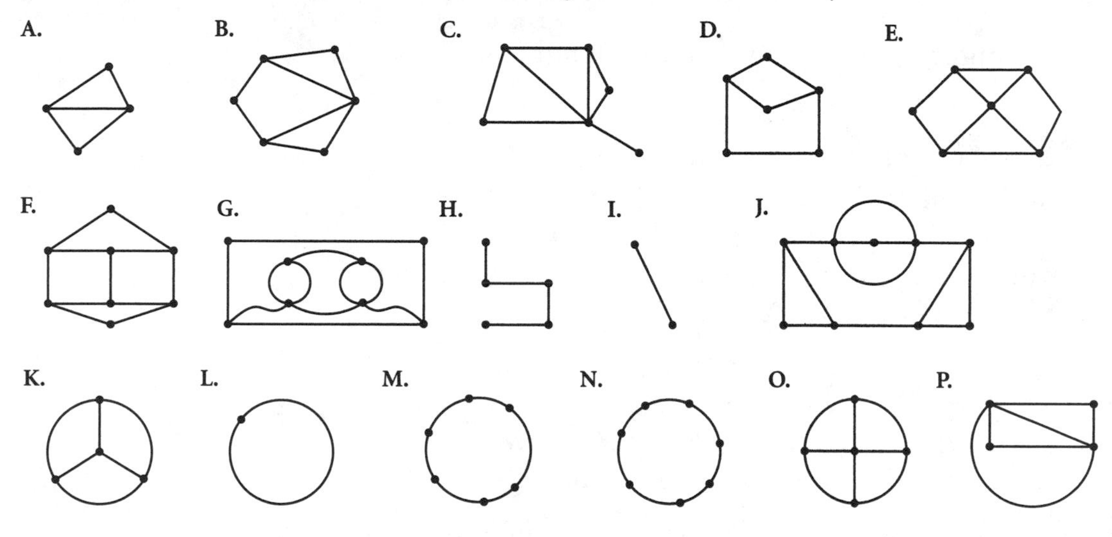

EXPLORATION

Map Coloring

In this activity you'll discover patterns in the number of colors needed to color a map.

Part 1: Coloring the Map of the United States

A *cartographer* is someone who makes maps. Suppose you are a cartographer and your job is to color a map of all the states in the United States. To make each state easy to distinguish from the others, states that share a border cannot be the same color, but states that touch only at a corner point can be the same color. Follow these rules to color the map of the United States provided by your teacher. As you color the map, determine your best answers to these two questions:

1. What is the least number of colors you need to color the map of the United States?
2. What is the least number of colors you need to color any map of any part of Earth?

Part 2: Coloring Two Special Kinds of Maps

The least number of colors needed to color a map is called the *chromatic number* of the map. (The prefix *chroma-* is from the Greek word for color.) In this section you will invent two types of maps and find their chromatic numbers.

1. On an unlined sheet of paper, randomly draw 10 to 15 straight intersecting lines that go all the way across the paper. Assume the different regions are countries on a map. Color the map with the least number of colors possible. Record the chromatic number of the map.
2. Tie a 1-meter-long piece of string into a single loop. Drop it onto a piece of plain paper. Use just enough cellophane tape to hold the string in place. Again, think of the different regions as countries on a map. Your map consists only of the regions enclosed by string. Color the map with the least number of colors possible. Record the chromatic number of the map.

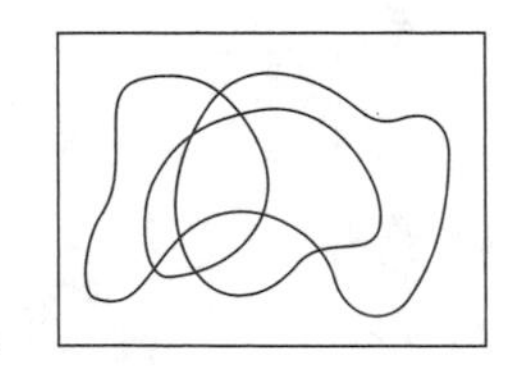

3. Compare your answers to the last two questions with those of others who have made the same kinds of maps. If you are working by yourself, make a few more maps using the same two techniques. Record any patterns you observe between the different kinds of maps, and try to explain them.

(continued)

Part 3: Finding the Largest Chromatic Number of Any Map

In this section you will explore in more detail the largest number of colors you would ever need if you were a cartographer coloring maps of Earth.

1. Design a simple map of three regions that cannot be colored with only two colors.
2. Design a simple map of four regions that cannot be colored with only three colors.
3. Try to design a map that requires more than four colors. Describe your results.

Part 4: The History of the Four-Color Theorem

Cartographers have known for centuries that they never need more than four colors to color any map (although they often use more than four colors for convenience). Mathematicians, however, have had a difficult time proving this is true. Research the history of the famous Four-Color Theorem and write a summary of your findings. Include a discussion of the controversial computer proof written in 1976 by the mathematicians K. Appel and W. Haken.

Part 5: Maps on the Torus

The surface of a solid that looks like a donut or bagel is called a torus. Investigate the largest possible chromatic number of a map on a torus. (*Hint:* It is more than four!)

EXPLORATION

Map Coloring

Map of the United States

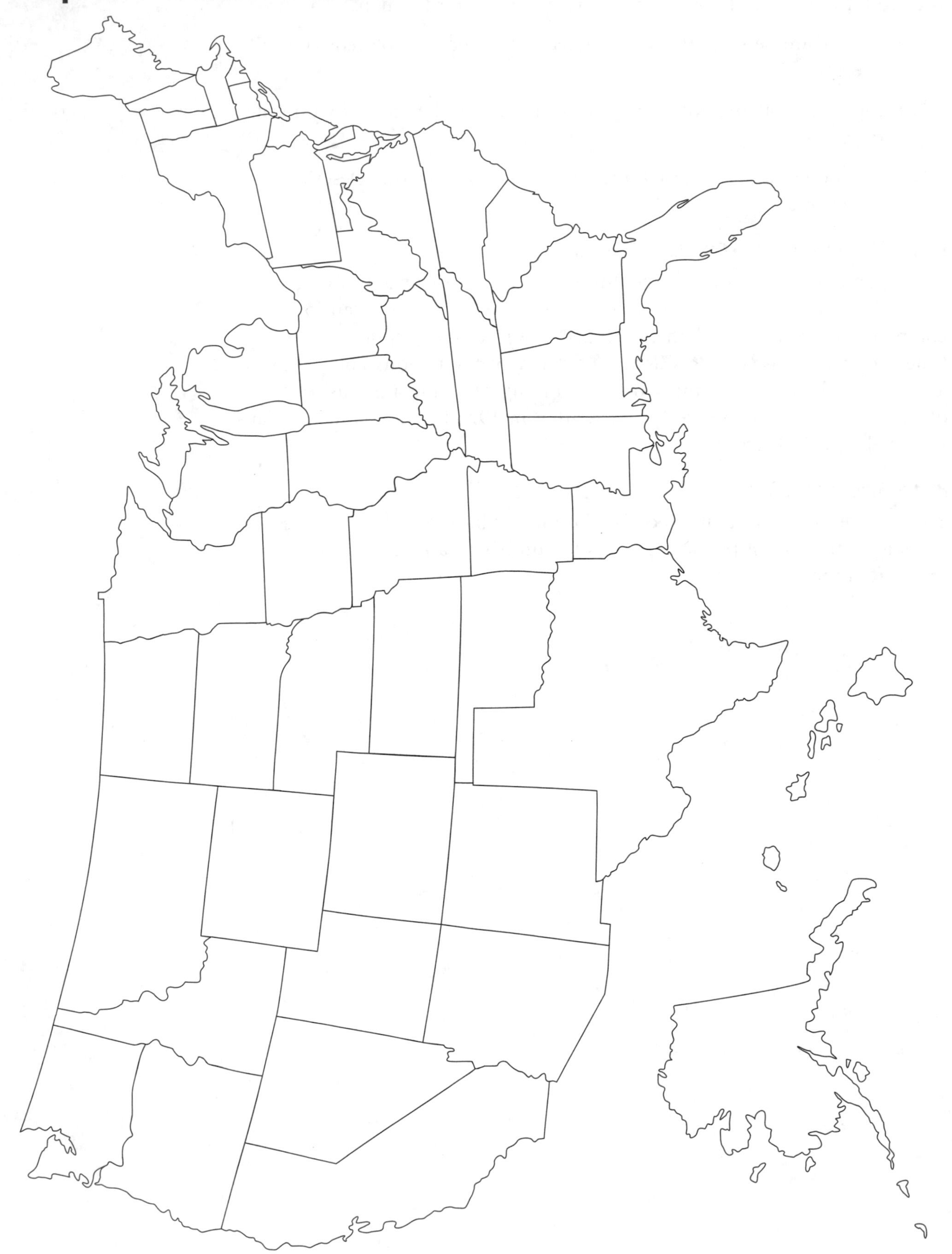

PROJECT • Sol LeWitt

Students follow a detailed set of instructions to construct a large work of art in the manner of Sol LeWitt. If students write their own sets of instructions, you might then have them construct each other's artworks. If students don't have a wall available on which to create their drawings, you could cover one with butcher paper, or they could use poster board. You might use this project with Lessons 3.1–3.5.

Outcome

- Student correctly constructs Michael Serra's *Wall Drawing #1* or creates his or her own wall drawing and gives a correct set of instructions for it.

PROJECT • In Perspective

This is a collection of four short projects that you can use as extensions of the Exploration Perspective Drawing. Students can also use geometry software to complete these projects.

Outcome

- Student correctly renders a perspective drawing.

Extra Credit

- Student combines two or more techniques in the drawing.

PROJECT

Sol LeWitt

Contemporary American artist Sol LeWitt (b 1928) is considered one of the pioneers of conceptual art. What is conceptual art? In Sol LeWitt's own words,

> In conceptual art the idea or concept is the most important aspect of the work. When an artist uses a conceptual form of art, it means that all of the planning and decisions are made beforehand and the execution is a perfunctory affair. The idea becomes a machine that makes the art.

LeWitt's wall drawings can be completed by LeWitt himself, by a LeWitt-trained team of assistants, or sometimes even by a group of local volunteers or the owner of the artwork. LeWitt has compared his instructions to a musical score and the first drawing of the art piece to the premier performance of the musical composition. Look on page 61 in your book to see an example of a Sol LeWitt drawing.

For this project, you can either create your own wall drawing or use Michael Serra's *Wall Drawing #1* below. If you create your own drawing, be sure to include a detailed set of instructions that someone else can follow. You may, without having to purchase Michael Serra's art, create *Wall Drawing #1* on a rectangular wall at your school (if done as a school project) or on a rectangular wall in your home (with your parent's permission, of course!). If you believe you are successful in your interpretation of the instructions, send a photograph of the work on the wall showing the complete drawing to Michael Serra, c/o Key Curriculum Press, 1150 65th St., Emeryville, CA 94608, and he will send you a Certificate of Authenticity.

Wall Drawing #1

Label the rectangular wall $ACEG$ with $\overline{AC}$ the bottom edge with the floor and $\overline{EG}$ the edge with the ceiling, $\overline{AG}$ the left edge and $\overline{CE}$ the right edge. Locate three points on each side of the rectangle so that each side is divided into four congruent segments. As you move counterclockwise about the rectangle, label the points in the following order: A, B_1, B_2, B_3, C, D_1, D_2, D_3, E, F_1, F_2, F_3, G, H_1, H_2, H_3.

Invisible parallel lines H_3D_1 and H_1D_3 intersect invisible parallel lines F_3B_1 and F_1B_3 at four points: I, J, K, and L. These four points determine four congruent rectangles: AB_1IH_3, CD_1JB_3, EF_1KD_3, and GH_1LF_3. Construct the four rectangles.

(continued)

Divide each of the four congruent rectangles into four congruent rectangles by segments connecting the midpoints of the opposite sides. Let M be the midpoint of $\overline{LF_3}$, N be the midpoint of $\overline{LH_1}$, O be the midpoint of $\overline{IH_3}$, P be the midpoint of $\overline{IB_1}$, Q be the midpoint of $\overline{JB_3}$, R be the midpoint of $\overline{JD_1}$, S be the midpoint of $\overline{KD_3}$, and T be the midpoint of $\overline{KF_1}$. Do not construct rectangle $LIJK$, but locate midpoints U, V, W, and X, where U is the midpoint of $\overline{LK}$, V is the midpoint of $\overline{LI}$, W is the midpoint of $\overline{IJ}$, and X is the midpoint of $\overline{JK}$.

With V as center and $\overline{B_1F_3}$ as diameter, construct a semicircle to the right of the diameter. With V as center and $\overline{MP}$ as diameter, construct a semicircle to the right of the diameter. With W as center and $\overline{IJ}$ as diameter, construct a semicircle above the diameter. With W as center and $\overline{OR}$ as diameter, construct a semicircle above the diameter. With X as center and $\overline{TQ}$ as diameter, construct a semicircle to the left of the diameter. With X as center and $\overline{F_1B_3}$ as diameter, construct a semicircle to the left of the diameter. With U as center and $\overline{LK}$ as diameter, construct a semicircle below the diameter. With U as center and $\overline{NS}$ as diameter, construct a semicircle below the diameter.

Select four different-color paints of your choice. Paint each of the four rectangles in rectangle H_1LF_3G one of the four chosen colors. Reflect the four colored rectangles over vertical line UW so that the paint color in the smaller upper-left rectangle in rectangle H_1LF_3G is the same color as the smaller upper-right rectangle in rectangle D_3KF_1E. Reflect the two painted upper rectangles H_1LF_3G and D_3KF_1E over horizontal line VX so that the paint color in the smaller upper-left rectangle in rectangle H_1LF_3G and the smaller upper-right rectangle in rectangle D_3KF_1E is the same color as the smaller lower-left rectangle in rectangle AB_1IH_3 and the smaller lower-right rectangle in rectangle CD_1JB_3. Shade in the remaining regions so that no two regions sharing an edge have the same color. Regions sharing only a point may have the same color.

PROJECT
In Perspective

High-Rise Complex

Skyscrapers are challenging to draw in perspective because they can be made of many different rectangular solids and thus have many different vanishing lines. Follow these steps to draw a block of skyscrapers in two-point perspective.

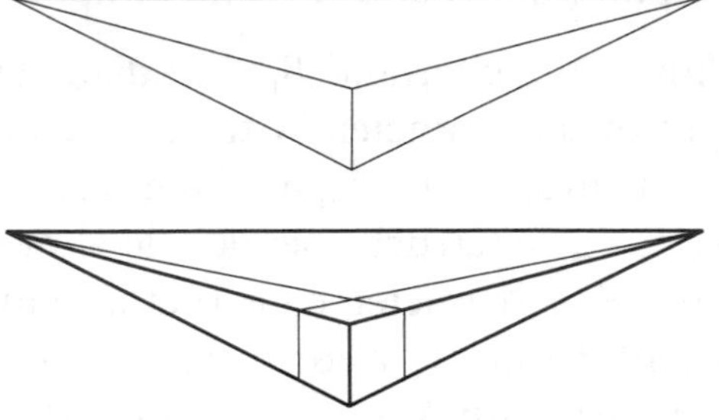

Step 1 Begin with a horizon line and two vanishing points. Draw the front vertical edge of your building with all the vanishing lines.

Step 2 Complete the two-point perspective view of the first building.

Step 3 Draw in a couple of the taller buildings. Start with the front vertical edge of each building and draw the vanishing lines. Complete the perspective view.

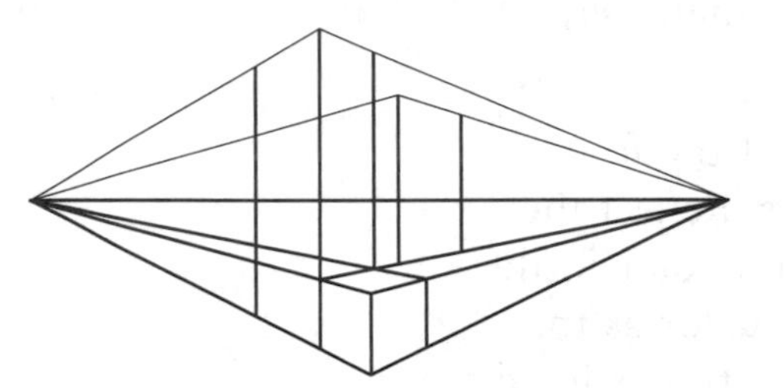

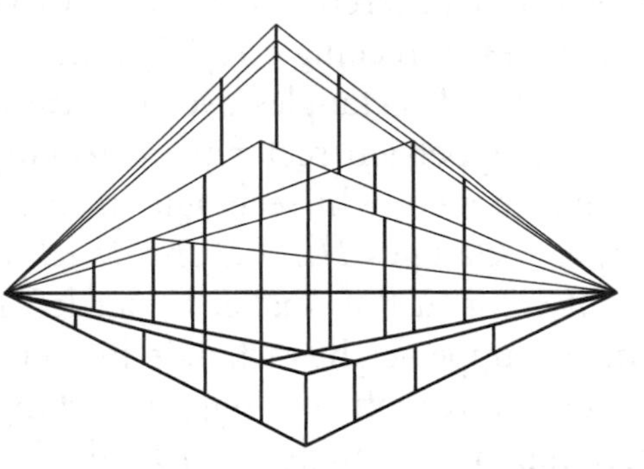

Step 4 Create additional buildings and use vanishing lines to add architectural details.

Step 5 Erase all unnecessary lines and add other details.

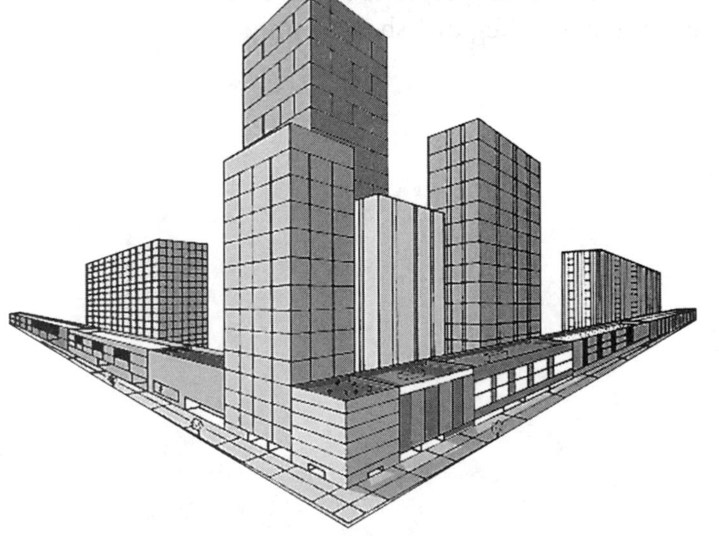

Can you find a parallelogram in your drawing? A trapezoid? A triangle? What would these shapes represent if these were real buildings?

PROJECT
In Perspective

Block Lettering

You can use perspective drawing to create letters or words that appear three-dimensional. This is useful for giving emphasis to an element of a design. Follow these steps to draw letters or words with one-point and two-point perspectives.

Drawing Block Letters in One-Point Perspective

Step 1 Draw a word in block letters. You might draw your name or initials. Draw a horizon line parallel to the bottom edge of your word. Select a vanishing point on the horizon line.

Step 2 Draw vanishing lines from all corner points of the block letters back to the vanishing point. Select a thickness for your block letters and draw line ℓ parallel to the horizon line.

Step 1

Step 2

Step 3 To create the back edges of your letters, draw lines parallel to the front edges, starting and ending on the points where line ℓ intersects the vanishing lines.

Step 4 Erase all the extra lines and shade all the sides and tops of the letters.

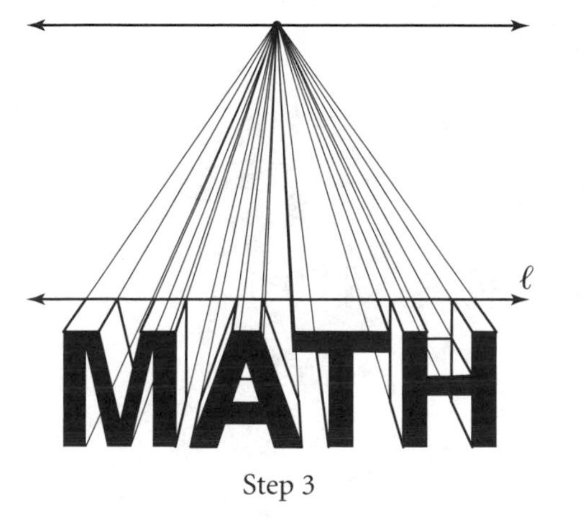

Step 3

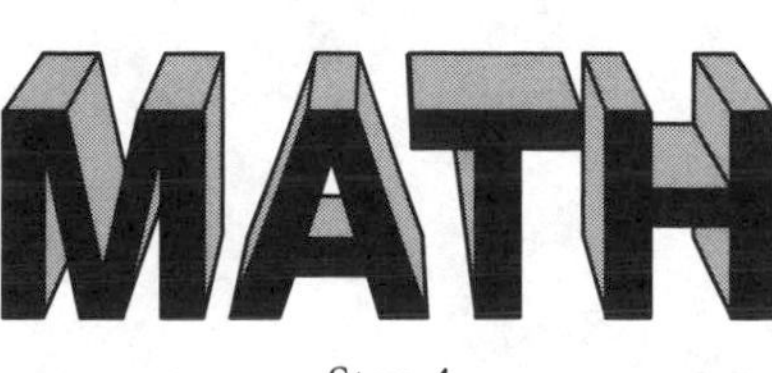

Step 4

(continued)

Drawing Block Letters in Two-Point Perspective

Step 1 Draw a box in two-point perspective. Label the points as shown. Make the height *CD* and the width *CE* of your box about the same, 6 cm, for example. Do not erase the vanishing lines. Your first letter will fill the front right face of this box.

Step 2 Select a distance between the first and second letters by drawing vertical segment *GH*. If you used 6 cm for the width *CE*, use about 1 cm for *EG*. Select a width for your second box by drawing vertical segment *IJ*. This box will eventually house your second letter. If you used 6 cm for *CE*, use about 3 cm for *GI*. Next, select a width for the space between the second and third boxes by drawing vertical segment *KL*. If you used 1 cm for *EG*, use 0.5 cm for *IK*. Repeat this procedure for the third box, and so on.

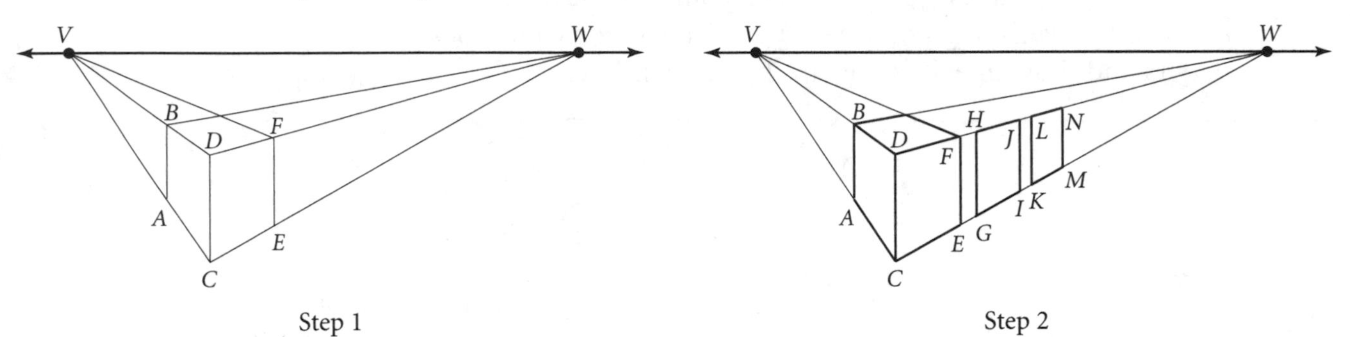

Step 1 Step 2

Step 3 Design a letter on the front face of each box. Draw diagonal segments *CF*, *DE*, *HI*, *GJ*, *LM*, and *KN*. The points where these diagonals intersect are the perspective centers for each front face. Draw vertical segments through these centers. Label the center in the first box *P*. Draw line *PW*. Use this line to center each block letter on its front face.

Step 4 Draw all the top vanishing lines from the top front corners to the back edges of the solid letters. Draw all the vertical edges at the backs of the solid letters. Draw all the remaining vanishing lines. With a pen or felt tip marker, outline all the edges of the solid letters. Erase all other lines. Decorate your design.

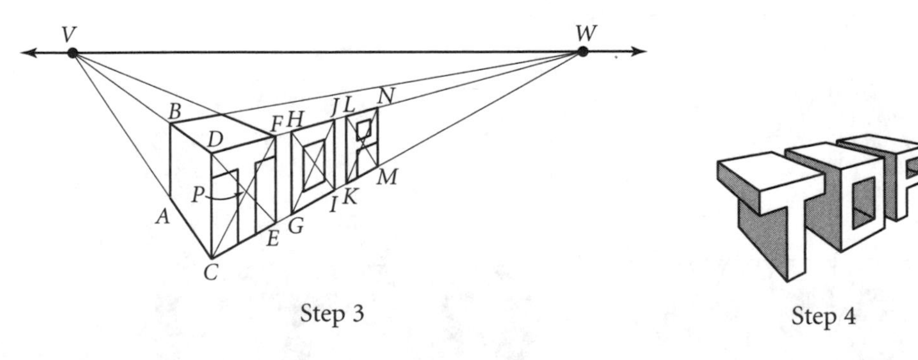

Step 3 Step 4

PROJECT

In Perspective

A Tiled Floor

Follow these steps to construct a perspective view of a tiled floor.

Step 1 Draw a pair of horizontal parallel lines. The top line is your horizon line. The lower line will become the front edge of your first row of squares. On the lower line, mark off eight equal lengths. On the top line (your horizon line), select a vanishing point. Draw all nine vanishing lines. Draw a line parallel to line *AB* and the horizon line. This line determines the back edge of your first row of squares.

Step 2 Draw diagonal segments *BD* and *AC*, and extend them to the horizon line. Diagonal segments *BV* and *AW* should cross the center vanishing line at the same point.

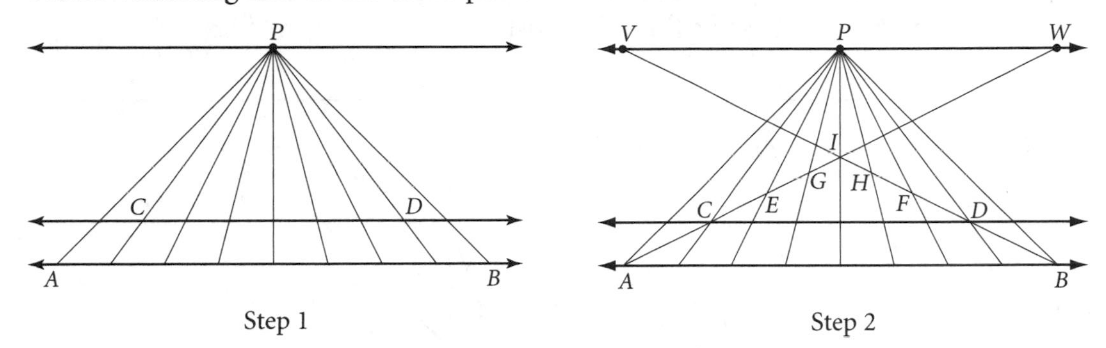

Step 1 Step 2

Step 3 Mark off points where diagonal segments *BV* and *AW* intersect the nine vanishing lines. Draw a line through points *E* and *F*. This is the back line for the second row of squares. Draw a line through points *G* and *H*. This is the back line for the third row. Continue in this fashion, drawing lines through pairs of points until all nine horizontal lines have been drawn.

Step 4 Shade alternating squares. Erase all unnecessary portions of horizontal and vanishing lines. For a special effect, you might add a slim rectangle to the front of the tile pattern to give it thickness.

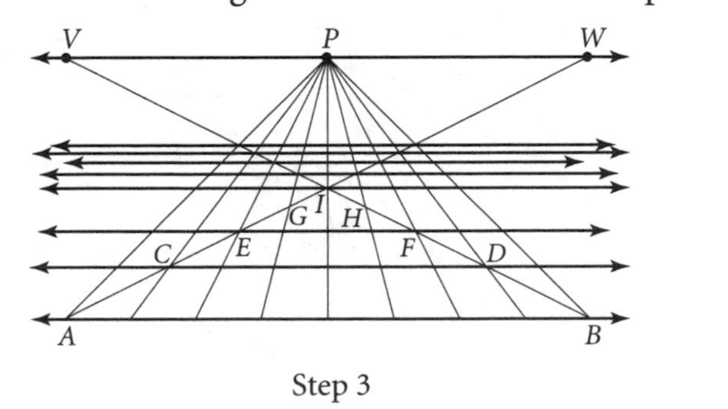

Step 3

Step 4

PROJECT
In Perspective

Spacing Fenceposts

Follow these steps to draw fenceposts in perspective.

Step 1 Start with a horizon line and vanishing point *V*. Draw the fencepost nearest the viewer and label it segment *AB*. Draw vanishing lines from the top and bottom of the post. Draw the second post parallel to the first at a distance that looks pleasing to you. Draw diagonal segments *AD* and *BC*. The diagonals intersect at point *P*.

Step 2 Draw a line connecting point *P* and the vanishing point. This line will pass through the centers of all the posts. Draw a line from point *B* through point *E*, the center of the second post. Extend this line until it meets the bottom vanishing line *AV*. Call this point *F*. Point *F* is the base of the third post.

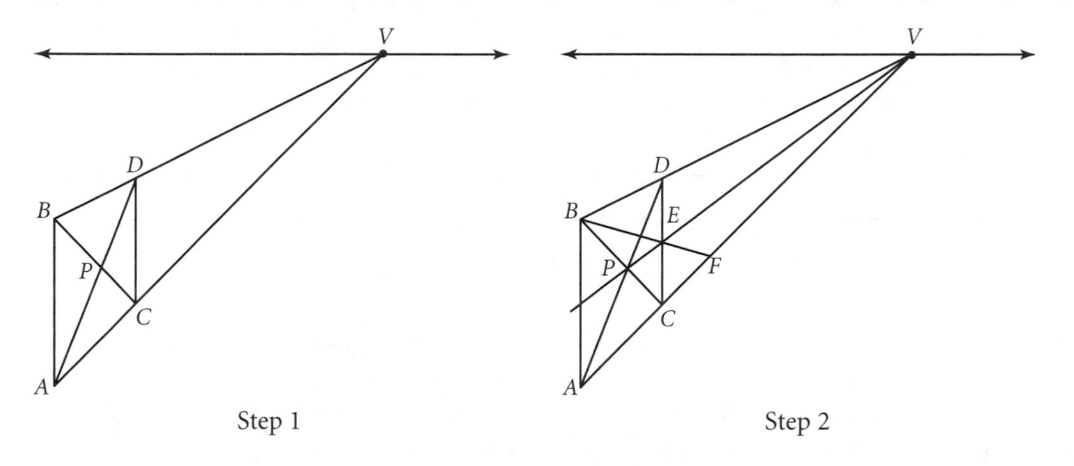

Step 1 Step 2

Step 3 Draw a line through points *D* and *G*, where point *G* is the middle point of the third post. Extend this line until it intersects the vanishing line *AV* at a point called *H*. This new point is the base of the fourth post. Continue until you have all the posts you want.

Step 4 Use these lines as a guide to draw the complete fence.

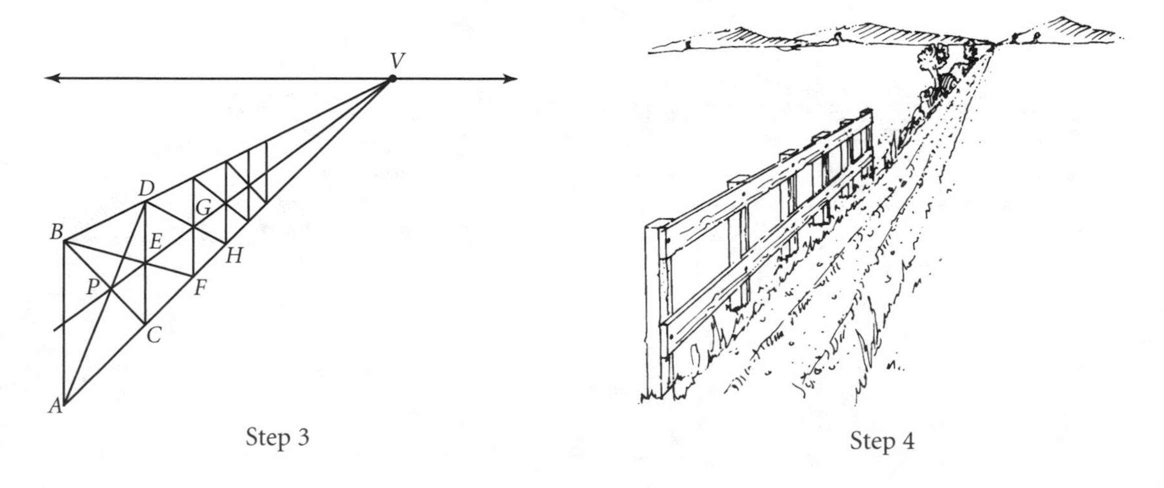

Step 3 Step 4

CHAPTER 4 TEACHER'S NOTES

EXPLORATION • Triangles at Work

Students model mechanical devices based on triangles. This activity can supplement Lesson 4.3.

MATERIALS

- cardboard strips with paper fasteners or small wood strips with nuts and bolts

GUIDING THE ACTIVITY

Answers will vary depending on the length of the sticks used. Answers are given for sticks 11 units long.

Step 2:

x	2	3	4	5	6	7	8	9	10
$m\angle A$	10°	16°	21°	26°	32°	37°	43°	48°	54°

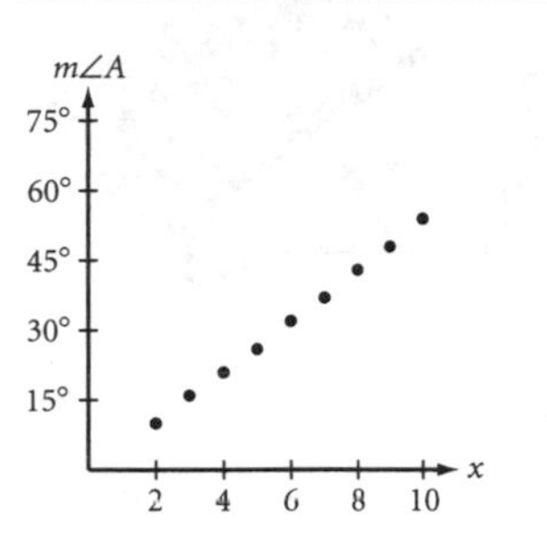

Step 3: 180°; 22 units

Step 4:

a. 8.4 units **b.** 3.8 units **c.** 0 units

Step 5: No, at a length of 4 units $m\angle A \approx 21°$.

Step 7:

AC	7.7	6.9	5.7	4.0	2.1
$m\angle A$	15°	30°	45°	60°	75°

Step 8: 7.5 units; 6.1 units

Step 9: If the angle changes at regular intervals, then the length of one side does not change at regular intervals, and vice versa.

Steps 10–12: Results will vary.

PROJECT • Buried Treasure

Students use triangle congruence conjectures to locate buried treasure on a map. This activity can supplement Lessons 4.4 and 4.5.

OUTCOMES

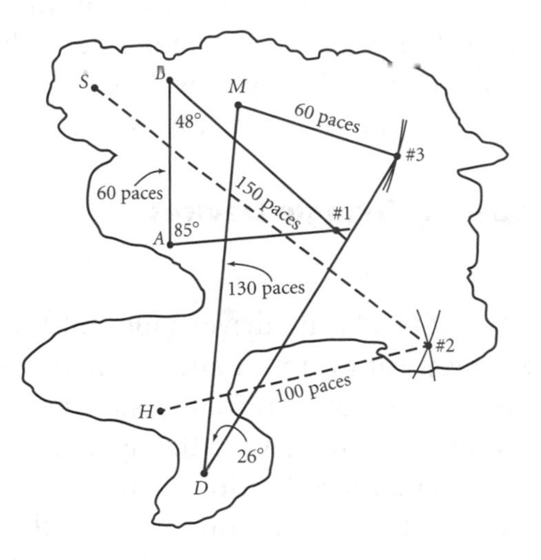

1. Yes, ASA **2.** Yes, SSS

3. No. SSA gives two possible locations for the treasure.

Student draws a map and uses a congruence conjecture to locate a buried treasure.

EXPLORATION • The Plumb-Line Method

Students use plumb lines to locate the center of mass of a triangle, then extend the method to quadrilaterals. This activity can supplement Chapter 4.

MATERIALS

- cardboard triangles
- pins
- string
- paper clips

GUIDING THE ACTIVITY

Step 3: The new line would intersect the others in the same point.

Step 5: The center of mass hangs directly below the pin.

Step 6: Sample answer by construction: Construct one diagonal of the quadrilateral. Locate the centroids of the two triangles formed by the diagonal, and label them C_1 and C_2. Repeat with the quadrilateral's second diagonal to find C_3 and C_4. Construct $\overline{C_1C_2}$ and $\overline{C_3C_4}$. These segments intersect at the centroid, or center of mass. The plumb-line method will also work, with no variation.

EXPLORATION
Triangles at Work

In Lesson 4.1, you learned that triangles are rigid and that, because of this property, they are used to add strength to structures. But triangles are also used in mechanisms that move by changing the length of one side. Cranes used for building are one example. Lengthening or shortening the top cable lowers or raises the boom.

In this activity you will investigate the properties of three mechanisms that take advantage of the rigidity of the triangle while allowing one side to vary in length: a dump truck, a reclining deck chair, and a car jack. To build models of these devices, you will need cardboard strips with paper fasteners or small wood strips with nuts and bolts.

Activity: Modeling Triangle Devices

Dump Truck

To unload a dump truck, the driver must tilt the bed of the truck from a horizontal position to an angle that will allow the contents to slide out the back of the truck. The steepness of the bed's angle ($\angle A$) depends on the length, x, of the hydraulic ram that extends to push up the bed. The more friction there is between the surface of the bed and the contents of the dump truck, the greater the measure of $\angle A$ must be before the contents will slide out. The driver can control how much the hydraulic ram extends (the value of x) and can stop the ram's movement when the truck's load slips out.

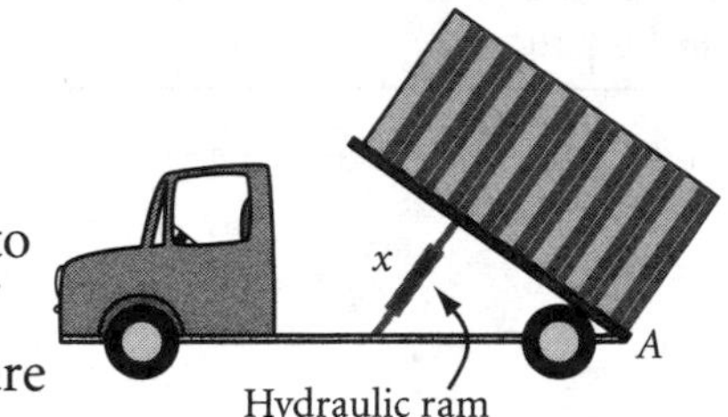

Step 1 To model the action of the hydraulic ram, set up three of your sticks with paper fasteners as shown at right.

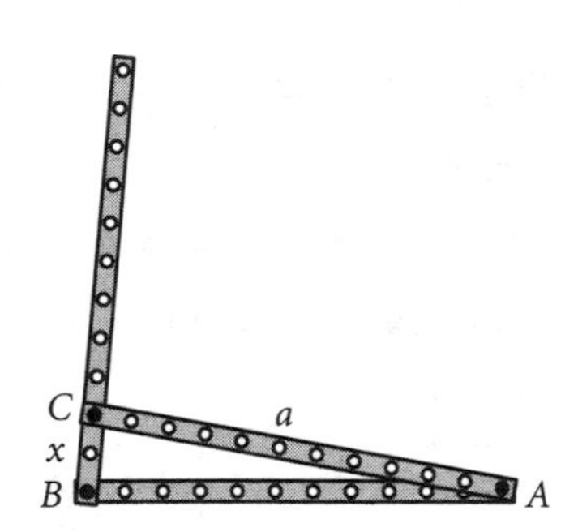

Step 2 Measure $\angle A$ with your protractor for each value of x (the number of spaces between your fasteners at points B and C), and enter your results in the table. Graph the points $(x, m\angle A)$ and draw a smooth curve connecting them.

x	2	3	4	5	6	7	8	9	10
$m\angle A$									

Step 3 What is the greatest possible value of $m\angle A$ for your model? What length should your hydraulic ram be to achieve that angle?

Step 4 About how long must the ram be extended to dump each load?

a. a load of bricks that will slide out when a 45° angle occurs

b. a load of sand that will slide out when a 20° angle occurs

c. water

Step 5 Suppose the ram can extend to a length of only 4 units. Can the dump truck successfully dump a load of wood that requires an angle of 30° to slide out? Explain.

(continued)

Reclining Deck Chair

Deck chairs are designed to support a person's upper body at different angles. To adjust the angle of the back of the chair, the person lifts the lower end (point C) of the support strut ($\overline{BC}$) out of the slot along the base of the recliner ($\overline{AC}$). It would be nice to have set positions along the base of the chair so that the back of the chair can be reclined at different angles in equal increments. Does this mean the slots along the base should be spaced equally?

Step 6 Use your sticks to model a chair with an adjustable reclining angle. In your model, let $AB = CB = 4$ units.

Step 7 Move point C and use your protractor to measure the reclining angle ($\angle A$) for the values in the table. What values of AC correspond to these angles? If you find the spacing is no longer equal, use a ruler and measure the length of x in centimeters for each angle. Make a graph of your results.

AC					
$m\angle A$	15°	30°	45°	60°	75°

Step 8 From your graph, what value of x would make an angle of 20°? 40°?

Step 9 Compare your results with your results for the dump truck model. Explain the difference between changing the angle at regular intervals and changing the length of one side at regular intervals.

Step 10 Find another tool or piece of equipment that uses a changing triangle to form a support. Make a diagram that explains the way it works, and model it with your sticks.

Car Jack

A car jack is used to raise one corner of a car, usually so that a tire can be changed. There are many different types of car jacks. The one shown at right is a rhombus with a variable diagonal that forms two triangles. The diagonal in the jack is a threaded bolt that can be made shorter or longer by turning a crank on one end of the bolt. Making the diagonal longer lowers the jack, and making it shorter raises the jack. Is there a linear relationship between the number of turns of the crank and the height of the jack?

Step 11 Use your sticks to model the situation (or, if you can, bring to class a car jack of the type shown). Close the car jack as far as possible (make the diagonal as long as possible).

Step 12 Record the height at 0 turns in a table. Give the handle 2 full turns to raise the jack. (If you're using sticks instead of a real jack, shorten the diagonal by 1 unit.) Record the new height. Give the handle another 2 full turns (or shorten the diagonal by 1 more unit), raising the jack further. Record the new height. Repeat until the jack is all the way up. Graph the results you recorded in the table. Is the relationship linear?

PROJECT

Buried Treasure

Captain Coldhart of the pirate ship *Xavier* always buries his treasures on Deadman's Island. Use this map of the island to find the locations of his buried treasures.

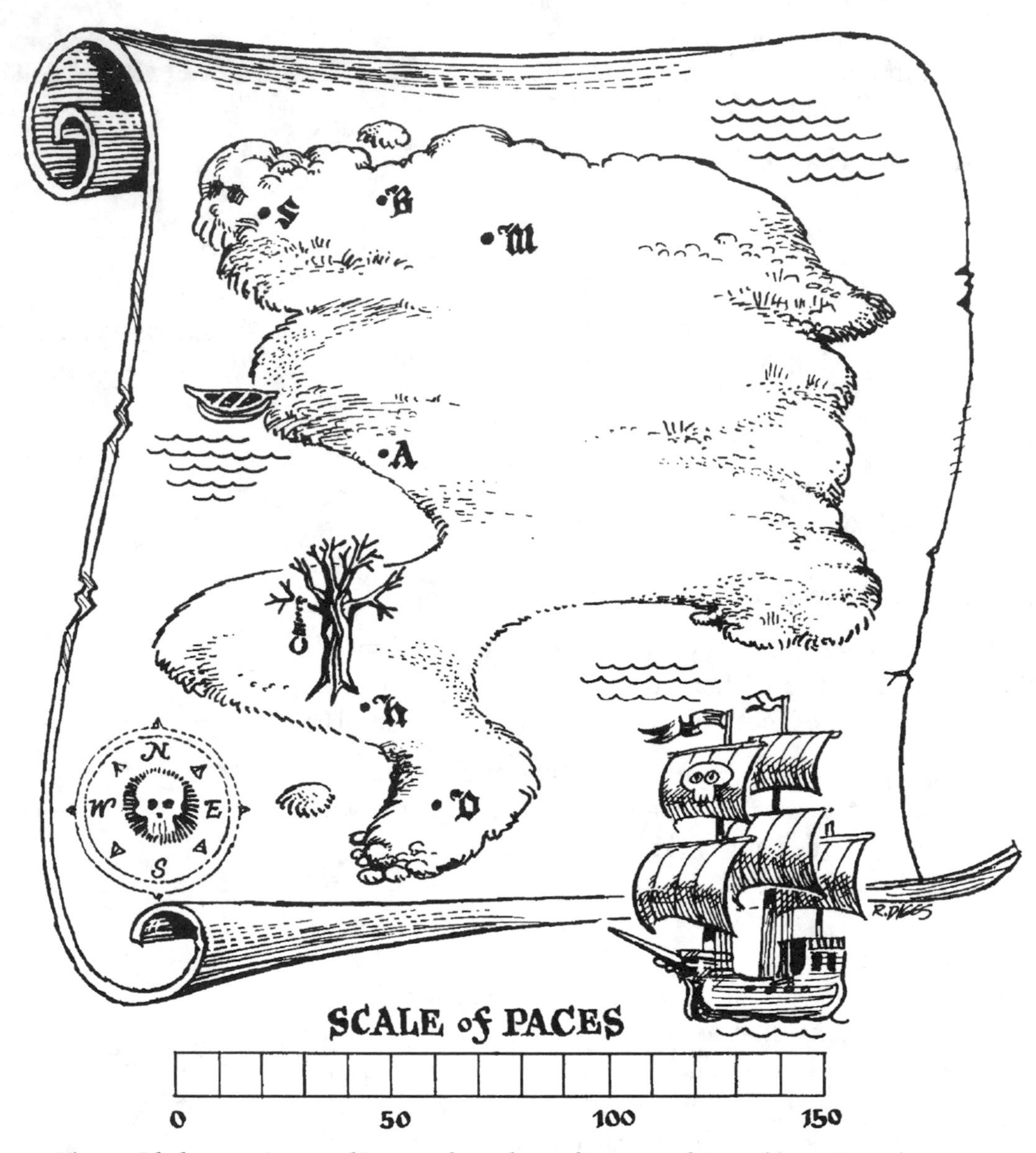

1. Pirate Alphonse is standing at the edge of Westend Bay (at point A), and his cohort, pirate Beaumont, is 60 paces to the north at point B. Each pirate can see Captain Coldhart off in an easterly direction, burying treasure. With his sextant, Alphonse measures the angle between the captain and Beaumont and finds that it measures 85°. With his sextant, Beaumont measures the angle between the captain and Alphonse and finds that it measures 48°. Alphonse and Beaumont mark their positions with large boulders and return to their ship, confident that they have enough information to return later and recover the treasure.

(continued)

Can they recover the treasure? Which congruence conjecture (SSS, SAS, ASA, or SAA) guarantees they'll be able to find it? If they can find it, use your geometry tools to locate the position of the treasure on the map. Mark it with an X.

2. Captain Coldhart, convinced someone in his crew stole his last treasure, has decided to be more careful about burying his current loot. He gives his trusted first mate, Dexter, two ropes, the lengths of which only the captain knows. The captain instructs Dexter to nail one end of the shorter rope to Hangman's Tree (at point H) and to secure the longer rope through the eyes of Skull Rock (at point S). The captain, with the ends of the two ropes in one hand and the treasure chest tucked under the other arm, walks away from the shore to the point where the two ropes become taut. He buries the treasure at the point where the two ropes come together, collects his ropes, and returns confidently to the ship.

   Has Captain Coldhart given himself enough information to recover the treasure? Which congruence conjecture (SSS, SAS, ASA, or SAA) ensures the uniqueness of the location? On your map, locate and mark the position of this second treasure if the two ropes are respectively 100 paces and 150 paces in length.

3. After the theft of two very important ropes from his locker and the subsequent disappearance of his trusted first mate, Dexter, Captain Coldhart is determined that no one will find the location of his latest buried treasure. The captain, with his new first mate, Endersby, walks out to Deadman's Point (at point D). The captain instructs his first mate to walk inland along a straight path for a distance of 130 paces. There, Endersby is to drive his sword into the ground for a marker (at point M), turn and face in the direction of the captain, turn at an arbitrary angle to the left, continue to walk for another 60 paces, stop, and wait for the captain. The captain measures the angle formed by the lines from Endersby to himself and from himself to the sword. The angle measures 26°. The captain places a boulder where he is standing (at point D), walks around the bay to Endersby, and instructs him to bury the treasure at this point.

   Has the captain given himself enough information to locate the treasure? If he has, determine the unique location of the treasure. If he does not have enough information, explain why not. How many possible locations for the treasure are there? Find them on the map.

Now make a map of your own. Identify necessary landmarks and write a story describing how you could locate a place to bury—and later find—a treasure. Use at least one triangle congruence conjecture in your method.

EXPLORATION

The Plumb-Line Method

In Lesson 3.8, you learned that the centroid of a triangle is the center of mass, and you constructed a triangle on cardboard and hopefully balanced it on its centroid. There is another way to locate the center of mass of a triangle, using a plumb line instead of a geometric construction. Whereas the geometric construction and patty-paper construction methods of finding the center of mass worked only for triangles, the plumb-line method works for any polygon. To make and use a plumb line, you will need a cardboard triangle, pins, string, and paper clips.

Activity: The Cardboard Triangle

Step 1 Stick a pin anywhere through the interior of the cardboard triangle. Pin the triangle loosely to a bulletin board. The triangle should hang freely by its own weight.

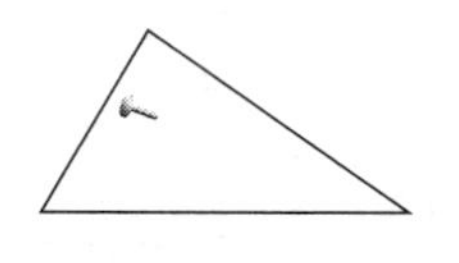

Step 2 Tie a paper clip to a string, and loop the string over the pin so that the weighted string hangs over the interior of the triangle. This string with a paper-clip weight is a **plumb line.** Draw a line on the cardboard triangle, showing the path of the string.

Step 3 Stick the pin through another point in the interior of the triangle and hang the plumb line. Draw another line on the cardboard triangle, showing the new path of the string. What do you think would happen if you repeated this step with yet another point?

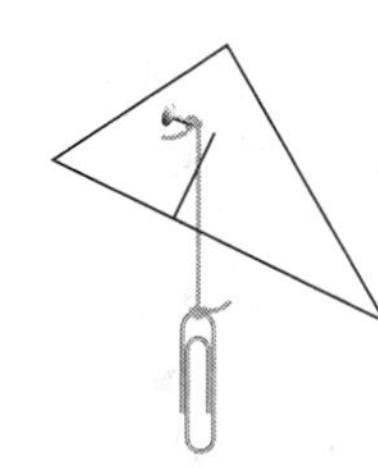

Step 4 The center of mass of the cardboard triangle is where the two lines cross. Check to see if your cardboard triangle balances on that point.

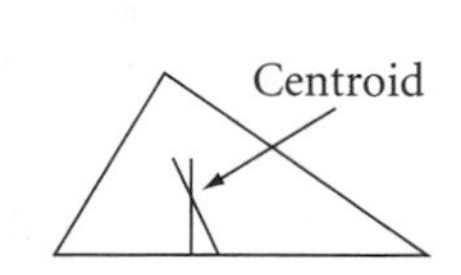

Step 5 Explain why this method works.

Step 6 Find a method for locating the center of mass of a quadrilateral. You may use a construction method or a plumb line. Explain why your method works.

EXPLORATION • Quadrilateral Linkages

Students model mechanical devices based on quadrilaterals, particularly the parallelogram. This activity can supplement Lesson 5.5. Part of Question 1 is Exercise 14 in that lesson.

MATERIALS

- Geostrips, cardboard strips with paper fasteners, or small wood strips with nuts and bolts
- string
- cellophane tape
- pop-up cards or books, *optional*

GUIDING THE ACTIVITY

Step 2: Yes, the parallelogram stays a parallelogram. No, the isosceles trapezoid doesn't even stay a trapezoid. Better names will vary.

Step 3: $\overline{CD}$ remains horizontal in the parallelogram. If you move to the right in the trapezoid, point C becomes higher than point D, and vice versa if you move to the left.

Step 4: In the parallelogram, point C moves in a circular path about center B with radius $\overline{BC}$. Point C can be made collinear with points A and B, at which time it will also be collinear with points A and D (because $\overline{AD}$ will coincide with $\overline{AB}$). In the trapezoid, point C moves in the same circular path about B. Point C can be made collinear with points A and B or with points A and D, but not at the same time.

Step 5: In both figures, point C moves in a circular path about point B with radius $\overline{BC}$, and point D moves in a circular path about point A with radius $\overline{CD}$. But the orientations of points C and D relative to each other are different in the two figures. In the parallelogram, $\overline{CD}$ is always parallel to $\overline{AB}$. In the trapezoid, point C is higher than point D as it moves to the right of the starting position shown and lower as it moves to the left of the starting position.

Step 6: The linkage most resembles the trapezoid linkage. When the wheels turn to either side, the wheel on the turning side is made to turn more than the other wheel.

QUESTIONS

1. The trapezoid linkage is used in the rocking horse so that the head and tail rock relative to each other. The parallelogram linkage is used in the sewing box so that the drawers remain parallel to each other and the ground as they are pulled out.
2. Answers will vary.
3. No, the seat doesn't always tilt at the same angle.

PROJECT • Symmetry in Snowflakes

contributed by Sharon Grand Taylor

Many students remember creating paper snowflakes when they were in elementary school, to decorate their classroom windows and to give as gifts. This project allows them to enjoy that process again while incorporating the geometry concepts and vocabulary they have been using in class, especially the idea of symmetry and the relationships between angle measures and polygon shapes. It is also an object lesson on the importance of listening to and writing instructions, and can foster cooperation among students as they share their understanding and creative ideas. In addition, this project is another way for students to link art and creativity to mathematics so that they become more aware of this influence in the world around them. This project also works well with Chapter 7.

If you do not have time to demonstrate how to fold polygons in class, as detailed in the instructions below, you might hand out the folding instructions to the students. Have them write a short summary of the geometric features of their snowflakes, including relevant angle measures and symmetries.

MATERIALS

- scissors for each student
- four sheets of unlined paper (or newsprint) for each student
- three to six pieces cut from sheets of colored or decorative tissue paper or gift wrap for each student
- trash can or trash bag for each group
- decorative materials, such as sequins and glitter, *optional*
- laminating machine with laminating film

TEACHER PREPARATION

1. Purchase packages of colored or decorative tissue paper; this usually comes with about ten sheets per package, with each sheet measuring about 24 in. by 36 in. Using a paper cutter or scissors, cut all ten sheets into rectangles approximately 9 in. by 12 in. Don't worry about cutting exactly, because the folding takes care of cutting errors.
2. Following the directions for folding, make patterns for each type of polygon from pieces of newsprint or plain paper. Mark the first fold with a "1" next to

the fold, the second fold with a "2," and so on so that you will have a handy guide to use when demonstrating for your students. Do not cut along the cut lines of your patterns; instead, mark them so that you can tell the students where to cut. Be sure to label each pattern with the type of polygon it produces.

3. Following your folding pattern for each type of polygon, cut snowflakes from tissue paper and laminate them to show as examples.

MAKING THE PATTERNS

Students should use the plain paper or newsprint to make folding patterns like the ones you made for demonstration purposes. They should make and number the creases, add notes to make the pattern easier to follow, mark the cut lines, and label the type of polygon each pattern produces.

1. Lead students through the folds for a pentagon, but don't tell them what shape will result. Stress the geometry vocabulary while folding, using the terms *bisect, right angle,* and so on. Have the students measure the smallest angle of the resulting triangle and record the results; ask them to predict what shape will result from opening up the triangle. You should actually cut the paper, but the students need to save their papers to use as patterns. Open the paper and ask questions as to what shape it is; what the angle measures should be; what the relationship is between the measured angle, the resulting central angle, and the number of sides; what types of symmetry there are; and so on.
2. Using additional sheets of paper, lead the students through the folding patterns for a hexagon and an octagon, using vocabulary and questions as in Step 1. Ask the students to predict what polygon will result from the folds, then make the cut to verify their predictions. They should give reasons for their predictions.
3. Have the students follow their own patterns to fold the hexagon. This time, they should cut along the cut line, but they should not open the paper yet. Tell them to cut out designs along any of the folds, the outside edges, and/or the center. After they've had several minutes to cut, they should open up the pieces of paper to see the "snowflakes" they have created. You should do this also. Tell the students that the more they cut away, the lacier their snowflakes will be.

COMPLETING THE PROJECT

The students will use the tissue paper to create a snowflake based on each of the three polygons demonstrated; they will do this outside of class. They may use their own tissue or gift-wrap paper, and they may decorate their snowflakes with sequins, glitter, or other materials. The snowflakes should be flattened by pressing them between heavy books a night or two before they are due.

Use a large folder or piece of poster board to collect the snowflakes and keep them flat. Then laminate them by using a laminating machine with rolls of film or by using a laminating press; even delicate snowflakes can be laminated if they've been pressed flat. Lamination makes the snowflakes durable for display purposes and also enhances their appearance. The students can now share their designs with the class, write journal entries about the geometry concepts illustrated, and make a classroom display on bulletin boards or in windows.

GAME • Matho

contributed by Masha Albrecht

Matho is a game based on Bingo. Rather than numbers and letters, students fill their boards with answers to review questions. It works well as a review lesson, but you can play it any time. The materials for this project include everything you need for a Chapter 5 Matho review game. You can make a similar game for any chapter by creating different Matho questions and answers.

MATERIALS

- one Matho Board worksheet per student
- Review Exercises transparencies
- Review Answers transparency
- silly prizes, *optional*

HOW TO PLAY

1. Give each student a blank Matho board. Display the answers on the overhead. Students fill in their Matho boards by randomly copying the 24 answers into the 24 spaces on their boards.
2. Show the review exercises on the overhead, one at a time. Students find and cross off the correct answer as they solve the exercise. Because this is a review day, it's probably worthwhile agreeing on the answer before you move on to the next exercise. Don't hesitate to stop and answer any review questions that students may have along the way. Emphasize that this game is not a race. Some students might simply wait until the answer is announced to cross it off their boards. If this becomes a problem, stop announcing the answers for a few exercises.
3. To get Matho, students must have an entire row, column, or diagonal crossed off. (The middle space is a free space.) As soon as they have done this, they yell, "Matho!" At this point, check that they have crossed off correct answers. If you don't remember which answers are already used up, don't worry; the other students will help you out.

4. Continue uncovering review exercises. The winners are the first three students to get Matho. Also count the last student to complete Matho before the end of class as a winner; otherwise, students may lose interest in the game after the first three winners. Most of these Chapter 5 review exercises have short answers, so you should be able to get through most of them in a class period. In games such as these, students will have almost the whole board filled in (obviously there will be lots of Mathos at that point). It might be unnecessary to give a prize to the last winner, because that would be practically everyone in the class.

5. Hand out silly prizes to the winners before the end of class.

EXPLORATION

Quadrilateral Linkages

How can the wheels of a vehicle turn smoothly, without sideways drag, if the steering mechanism regulates only one axle? In 1818, German engineer George Lenkensperger invented a device that solved this problem. German-British printer Rudolf Ackermann (1764–1834) patented this device for him in England. Of course, Lenkensperger and Ackermann worked with horse-drawn carriages then. Automobiles were not invented until about 70 years later, but this invention is still called the Ackermann steering linkage, and you can see it if you look carefully under a car.

Inner wheel traces a smaller circle, so it must be turned at a greater angle..

Center of turning circle

In this activity you will investigate the properties of quadrilateral linkages like the Ackermann steering linkage. You will need Geostrips, or cardboard strips with paper fasteners, or small wood strips with nuts and bolts to build each model. You will also need string and cellophane tape.

Activity: Quadrilaterals at Work

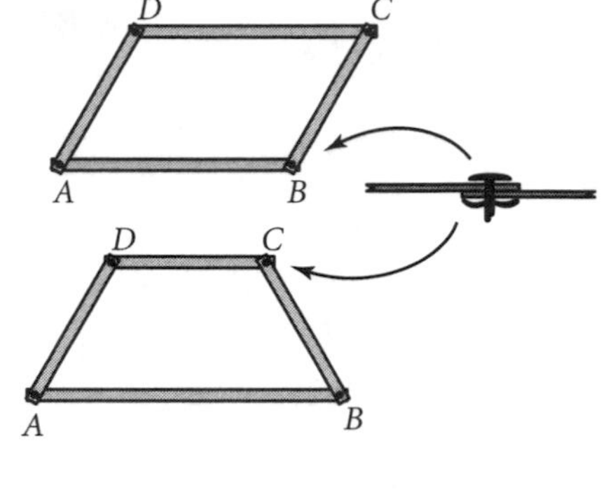

Step 1 Begin by building two linkages. At right is a parallelogram linkage, and below right is an isosceles trapezoid linkage.

Step 2 Does the parallelogram linkage stay a parallelogram when the linkage is moved around? Does the isosceles trapezoid linkage stay an isosceles trapezoid when the linkage is moved? What might be a better name for this quadrilateral?

Step 3 Set the parallelogram model vertically on your desk by holding the bottom bar $\overline{AB}$. Place the palm of your other hand lightly on the top bar $\overline{CD}$ and feel the motion of the bar as you move your hand back and forth. Do this with the isosceles trapezoid linkage, too. What is the difference in "feel" of the two sideways motions?

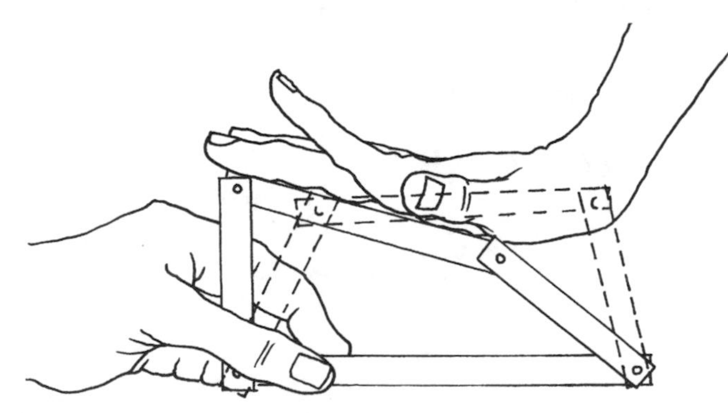

Step 4 Lay the parallelogram model flat on your desk, and hold or clamp the bottom bar $\overline{AB}$ to the desk so that it is fixed. Tug lightly on the top bar $\overline{CD}$ at point C. What happens? What path (locus of points) can point C take? Can you make point C collinear with points A and B? With points A and D? Now do this with the isosceles trapezoid linkage and answer the same questions.

(continued)

Exploration • Quadrilateral Linkages (continued)

Step 5 When the bar $\overline{AB}$ is fixed, what geometric figures describe the paths of points C and D for the parallelogram linkage? For the isosceles trapezoid linkage? Test your conjecture by temporarily replacing the paper fasteners at points C and D with pencil points to draw the paths.

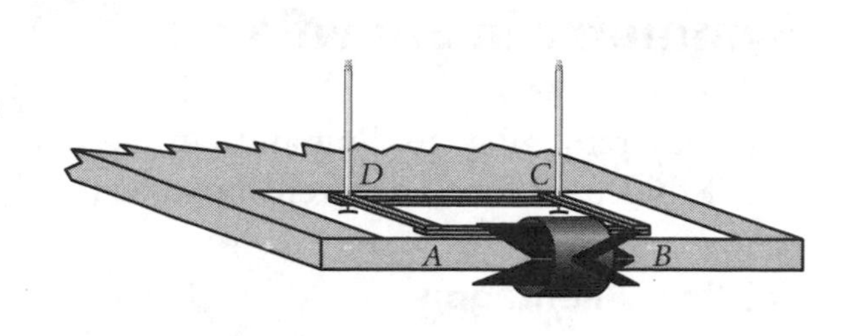

Step 6 The key to smooth turning is to have the wheels always facing at right angles to the line drawn from the center of the wheels to the center of the turning circle, as shown in the figure in the introduction. A top view of the linkage in a car's steering mechanism is shown at right. What type of linkage, parallelogram or trapezoid, does this linkage most resemble? Explain how this linkage is used to create a smooth-turning vehicle. To help explain the functioning of this mechanism, you might read about linkages in a book on mechanical design or auto repair.

Top view of steering linkage

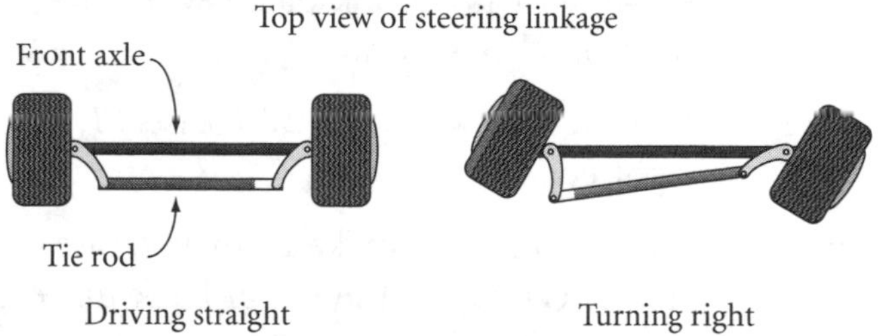

Questions

1. Quadrilateral linkages are used for many different mechanisms. Study the sewing box and rocking horse pictured here. Explain why a trapezoid linkage is used in the rocking horse and why a parallelogram linkage is used in the sewing box. What is the difference between the desired motions in these mechanisms?

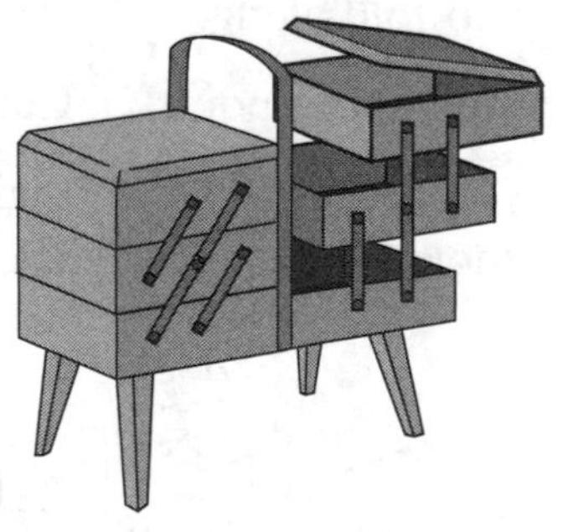

2. Pop-up cards are another example of parallelogram linkages. Find a pop-up card or book and explain how it was made.

3. Some kinds of patio chairs use a quadrilateral linkage like the one at right. The chains hanging from each arm are attached to the front and back of each end of the seat. The chains swing freely so that the seat can swing back and forth. Does the seat always tilt at the same angle from a horizontal position as it swings back and forth? Model the chair's chains with two pieces of string taped to the end of your desk so that they hang down. Tie them to the ends of a pencil to model one end of the seat.

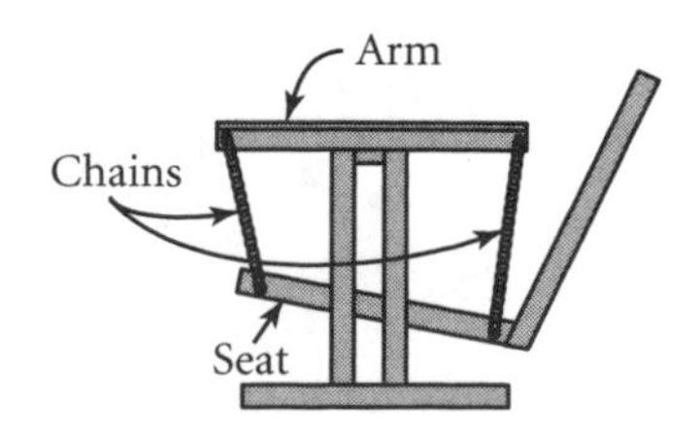

PROJECT
Symmetry in Snowflakes

You may remember folding and cutting paper snowflakes. In this project you'll fold paper into different polygons and use them to make snowflakes.

Folding a Pentagon

Step 1 Start with a piece of paper that is approximately 9 in. by 12 in. (a piece of notebook paper will work), and begin with paper in horizontal position.

Step 2 Fold in half with vertical crease *EF*, bringing left edge *AD* over to right edge *BC*.

Step 3 Fold *F* to *A* and make a small crease marking the midpoint of $\overline{FA}$ (point *G*). Open the paper back up to Step 2.

Step 4 Fold *E* to *G* and make crease *HJ*.

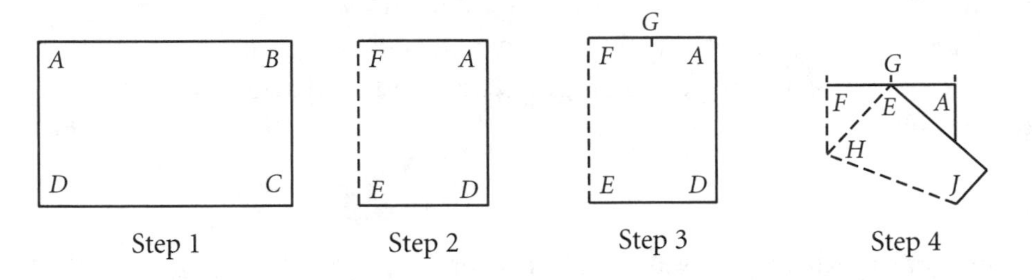

Step 1 Step 2 Step 3 Step 4

Step 5 Match $\overline{HJ}$ on top of $\overline{HE}$ and make crease *HK*. This crease should bisect $\angle JHE$.

Step 6 Fold the $\triangle HFG$ flap backward, under the paper, using crease *HK* as your guideline to fold along.

Step 7 Flip over by rotating about point *H*. Cut along $\overline{FG}$.

Step 8 Keep $\triangle HFG$. If you open it now, you will get a pentagon. If you want to make a snowflake, cut along the edges before opening.

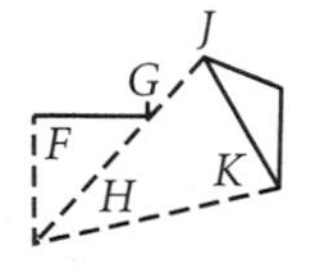

Step 5

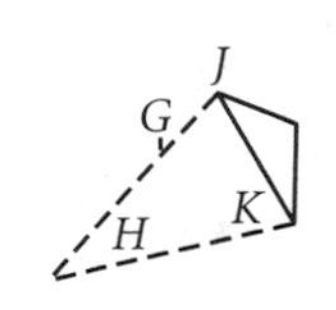

Step 6

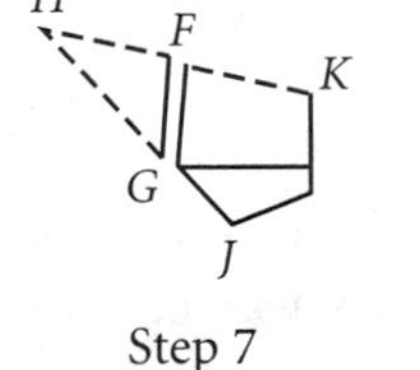

Step 7

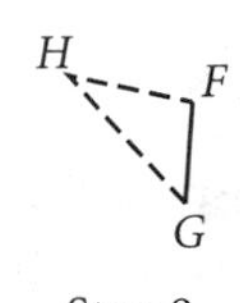

Step 8

(continued)

Folding a Hexagon

Step 1 Begin with paper in horizontal position.

Step 2 Fold in half with vertical crease *EF*, bringing left edge *AD* over to right edge *BC*.

Step 3 Fold in half with horizontal crease *GH*.

Step 4 Fold in half again, this time with horizontal crease *JK*.

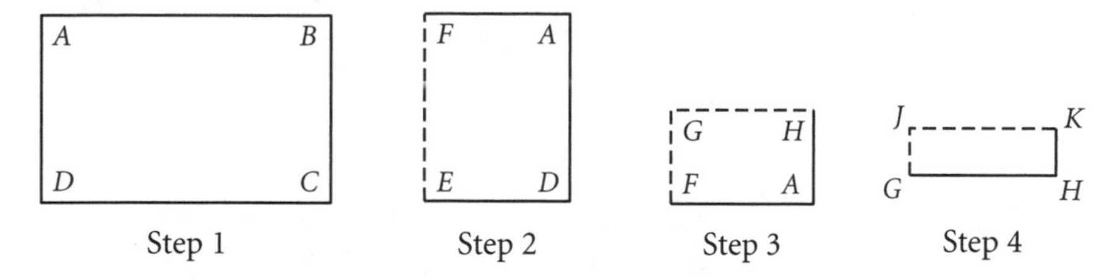

Step 5 Unfold once.

Step 6 Place lower-left corner *E*/*F* on crease *JK* so that you can make a crease through upper-right corner *G*. Make crease *GM*.

Step 7 Bisect $\angle MGH$ by folding under, using crease *GF* as your folding guide.

Step 8 Cut along $\overline{FM}$. Keep $\triangle GFM$. If you open it now, you will get a hexagon. If you want to make a snowflake, cut along the edges before opening.

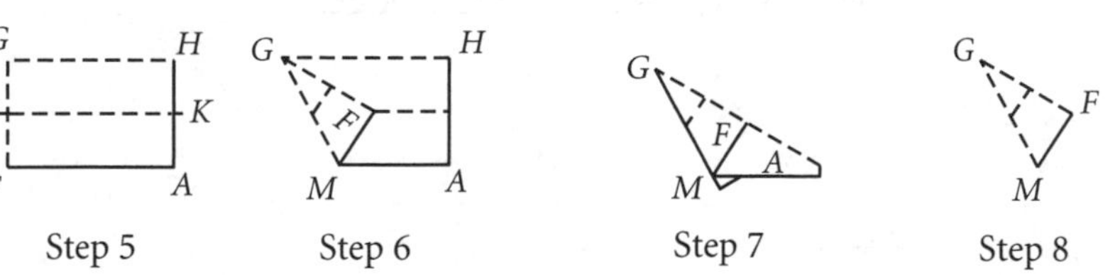

(continued)

Folding an Octagon

Step 1 Begin with paper in vertical position.

Step 2 Fold in half with horizontal crease *EF*, bringing top edge *AC* down to bottom edge *BD*.

Step 3 Fold in half with vertical crease *GH*.

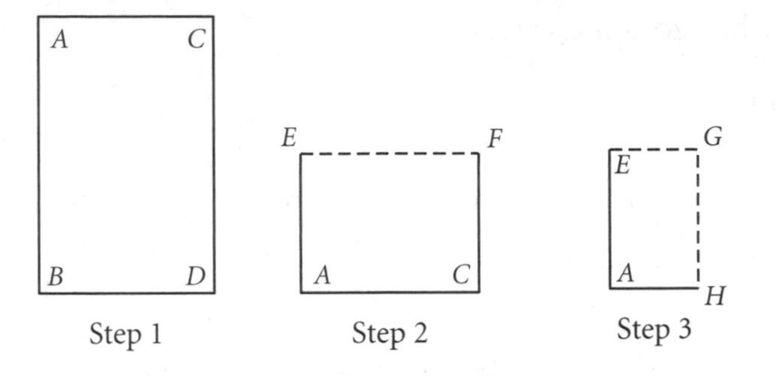

Step 1 Step 2 Step 3

Step 4 Bisect $\angle EGH$ by placing upper-left corner *E*/*F* on crease *GH* and making a crease through upper-left corner *G*. The new crease is *GJ*.

Step 5 Bisect $\angle JGF$ by placing edge *GF* on top of edge *GJ* and making a crease through point *G*. This crease is *GK*.

Step 6 Bisect $\angle JGH$ by folding under, using *GK* as your folding guide.

Step 7 Cut along *EK*. Keep $\triangle GEK$. If you open it now, you will get an octagon. If you want to make a snowflake, cut along the edges before opening.

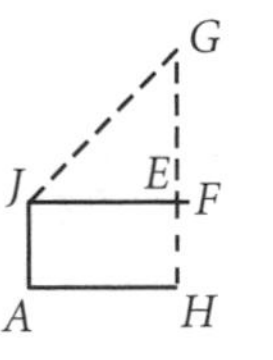

Step 4

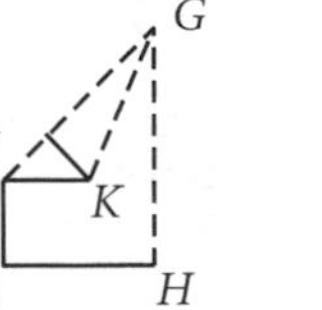

Step 5

Step 6

Step 7

GAME

Matho (page 1 of 4)

Matho Board

Matho

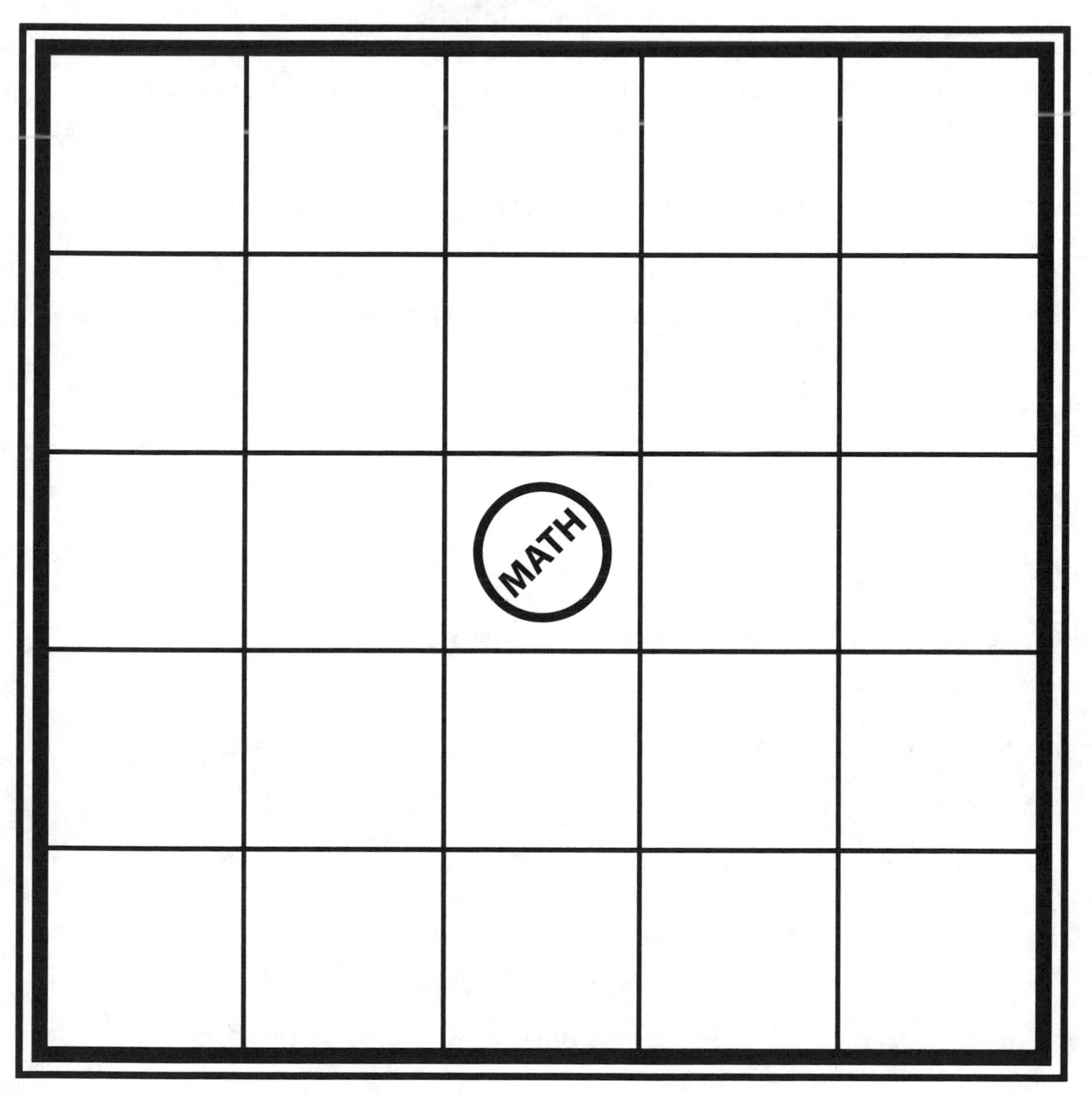

Chapter 5 Review Exercises

For Exercises 1–13, complete each statement using the best answer from your Matho Board.

1. A quadrilateral with four congruent sides is called a ____________.
2. The sum of the measures of the interior angles of a pentagon is ____________.
3. The sum of the measures of the exterior angles of a pentagon is ____________.
4. Any angle in a regular octagon measures ____________.
5. The measures of the interior angles of a(n) ____________ have a sum of 720°.
6. The nonvertex angles of a kite are ____________.
7. The diagonals of a kite are ____________.
8. The consecutive angles between the bases of a trapezoid are ____________.
9. The midsegment of a triangle is ____________ times the length of the base.
10. The midsegment of a trapezoid is the ____________ of the lengths of the bases.
11. A rhombus is a(n) ____________ with perpendicular diagonals.
12. A square is a(n) ____________ with congruent sides.
13. If a trapezoid has congruent base angles, it is ____________.

Matho (page 3 of 4)

Chapter 5 Review Exercises

For Exercises 14–19, answer each question and find the best answer on your Matho Board. You may want to sketch a picture first.

14. The midsegment of a trapezoid measures 3 cm. Find the length of one base if the other base measures 4.2 cm.

15. A student cuts a polygon along two lines and forms three congruent equilateral triangles. What shape was the original polygon?

16. A student reflects an obtuse triangle across the side opposite its obtuse angle. What quadrilateral consists of the original triangle plus the image triangle?

17. A kite has angles of measures 80° and 140°. Find the measure of another of its angles.

18. The measure of an exterior angle of a regular polygon is 30°. Find the number of sides of the polygon.

19. Describe polygon *ABCD* if *A* has coordinates (2, 1), *B* has coordinates (1, 2), *C* has coordinates (4, 5), and *D* has coordinates (4, 3).

In Exercises 20–24, find the measure of each angle in the diagram.

20. a

21. b

22. c

23. d

24. e

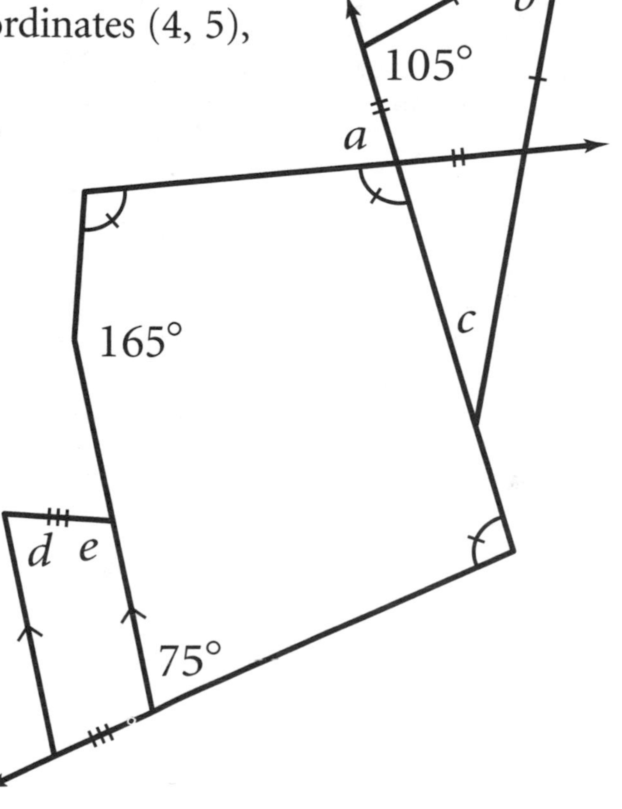

Matho (page 4 of 4)

Chapter 5 Review Answers

rhombus	perpendicular	rectangle	
hexagon	supplementary	75°	
congruent	average	1.8	60°
parallelogram	12	80°	
trapezoid	50°	25°	
trapezoid	135°	540°	0.5
kite	isosceles	360°	105°

EXPLORATION • Folding Paper Circles

contributed by Sharon Grand Taylor

Students explore properties of circles and triangles. You might use this activity as a midunit review after Lesson 6.3, as a review activity after completing Chapter 6, or as an informal assessment at the end of the chapter.

The activity introduces the tetrahedron in Step 8 and the truncated tetrahedron in Step 9, so you can also use it as a springboard for further study of Platonic and Archimedean solids. Once their folded circles are completed, students can use them for reference when they study similar figures. (Or, you could repeat the paper folding as a quick review.) This works especially well if students begin with circles of different sizes.

You could also give students the directions orally.

Materials

- plain paper
- compass
- scissors

Guiding the Activity

Step 2: Chord *AB* is halfway to the center. The folded part of the circle is the minor arc and the rest is the major arc.

Step 3: Chord *BC* is also halfway to the center. It has minor arc *BC* and major arc *BAC*. The lengths of chords *AB* and *BC* are equal.

Step 4: $\triangle ABC$ is equilateral because its sides are three congruent chords. The three minor arcs are congruent and have measure 120°. The measure of the angles of the triangle is half the measure of the arcs.

Step 5: *ABFE* is an isosceles trapezoid. $\overline{EF}$ is the midsegment of $\triangle ABC$, so it is parallel to $\overline{AB}$. $\angle A \cong \angle B$ because $\triangle ABC$ is equilateral.

Step 6: *ADFE* is a rhombus. $\overline{EF} \parallel \overline{AD}$ and $\overline{DF} \parallel \overline{AE}$ by the Triangle Midsegment Conjecture, and $\overline{AD} \cong \overline{DF} \cong \overline{FE} \cong \overline{EA}$ because each is half of one side of an equilateral triangle.

Step 7: Equilateral. Each of its side lengths is half the length of a side of $\triangle ABC$, which is equilateral. $\triangle DEF$ has $\frac{1}{4}$ the area of $\triangle ABC$ because four triangles the size of $\triangle DEF$ fit inside $\triangle ABC$.

Step 8: This solid has four sides that are four congruent equilateral triangles.

Step 9: This solid has five sides. Three sides are isosceles trapezoids $\frac{1}{4}$ the size of *ABFE*. One side is $\triangle DEF$ and one side is an equilateral triangle $\frac{1}{4}$ the size of $\triangle DEF$ (because it was constructed on $\triangle ADE$ [or $\triangle BDF$ or $\triangle CEF$] in the same way that $\triangle DEF$ was constructed on $\triangle ABC$, and $\triangle ADE \cong \triangle DEF$).

Step 10: The folded lines show equilateral triangles and isosceles trapezoids. Many geometry concepts are shown, including the Triangle Midsegment Conjecture. Geometry properties illustrated will vary.

PROJECT • The Art of Pi

Many works of art have been inspired by π. Students create their own artwork using π in some way. This project can supplement Lesson 6.5.

Outcome

- Student writes a poem, makes a collage, or in some way creates an artwork based on π.

PROJECT • Racetrack Geometry

Students use the properties of circles to design a fair racetrack. This is a guided version of the project on page 354 in Lesson 6.7 of the student book.

Outcome

Student designs a fair racetrack meeting the given specifications and provides supporting calculations. Sample track:

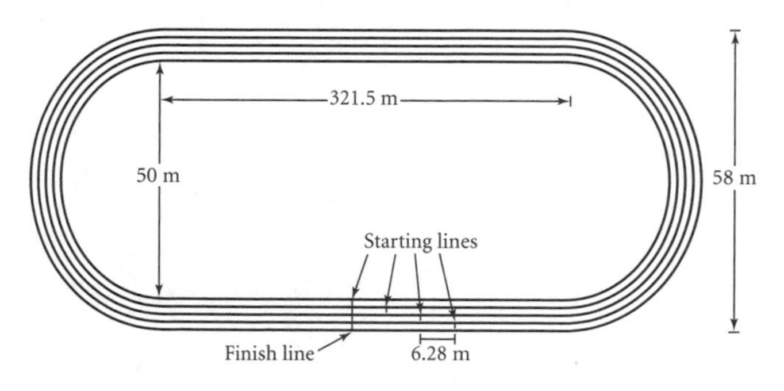

The radius of the circle does not play a part in determining the head start.

The width of the lane plays a part in determining the head start. The head start from lane to lane is always 2π times the lane width.

The lengths of the straightaways do not play a part in determining the head start.

Circular Track
Inner Radius 50 m, Lane Width 1 m

Lane	Radius (m)	Circumference (m)
1	50	100π
2	51	**102π**
3	**52**	**104π**
4	**53**	**106π**

Circular Track
Inner Radius 65 m, Lane Width 1 m

Lane	Radius (m)	Circumference (m)
1	65	130π
2	66	**132π**
3	**67**	**134π**
4	**68**	**136π**

Circular Track
Inner Radius 65 m, Lane Width 1.5 m

Lane	Radius (m)	Circumference (m)
1	65	130π
2	66.5	**133π**
3	**68**	**136π**
4	**69.5**	**139π**

Oval Track
Inner Radius 30 m, Straightaway 100 m, Lane Width 1 m

Lane	Radius (m)	Straightaway (m)	Total distance (m)
1	30	100	$200 + 60\pi$
2	31	100	**$200 + 62\pi$**
3	32	100	**$200 + 64\pi$**
4	33	100	**$200 + 66\pi$**

Oval Track
Inner Radius 30 m, Straightaway 200 m, Lane Width 1 m

Lane	Radius (m)	Straightaway (m)	Total distance (m)
1	30	200	$400 + 60\pi$
2	31	200	**$400 + 62\pi$**
3	32	200	**$400 + 64\pi$**
4	33	200	**$400 + 66\pi$**

Racetrack designs will vary depending on the width of the lanes. In all the designs the length of the straightaway must be about 321.5 m:

$$800 = 2x + 50\pi$$

$$800 - 50\pi = 2x$$

$$x = \frac{800 - 50\pi}{2} \approx 321.5$$

The length of the head start from lane to lane is 2π m, or about 6.28 m, times the lane width. One possible racetrack design based on a lane width of 1 m is shown on the previous page.

EXPLORATION

Folding Paper Circles

You know how to construct many geometric figures using a compass and straightedge or a square piece of patty paper. But what can you construct using a circle?

Activity: Triangles and Circles

Step 1 Cut a large circle from a piece of plain paper. Mark the center.

Step 2 Fold a point on the circle to the center. Call the crease segment *AB*. How far from the center is the chord? Identify the minor and major arcs for chord *AB*.

Step 3 Use point *B* as an endpoint of a second crease. Fold a second point of the circle to the center (segment *BC*). How far from the center is the chord? Identify the minor and major arcs for chord *BC*. What is true about the lengths of chords *AB* and *BC*?

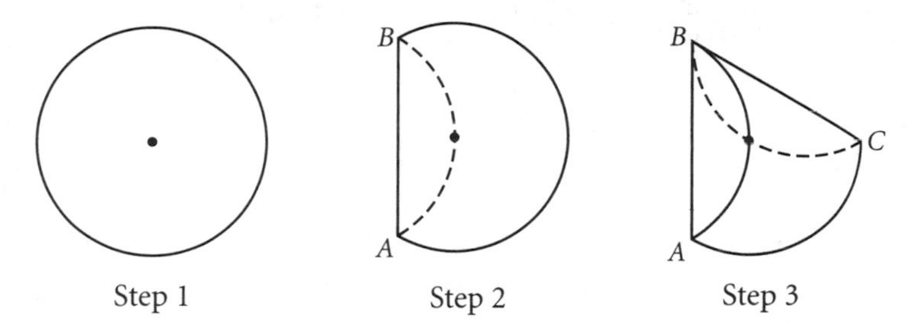

Step 1 Step 2 Step 3

Step 4 Fold segment *AC*. What kind of triangle is *ABC*? How do you know? Compare the three minor arcs. What are their measures? Compare the measures of the angles of the triangles to the measures of the arcs.

Step 5 Locate the midpoint of segment *AB* (point *D*). Fold point *C* to point *D*. Crease to form segment *EF*. What kind of quadrilateral is *ABFE*? Be as specific as possible. Explain how you know this.

Step 6 Fold point *B* across to point *E* along segment *DF*. What kind of quadrilateral is A*DFE*? Explain how you know this.

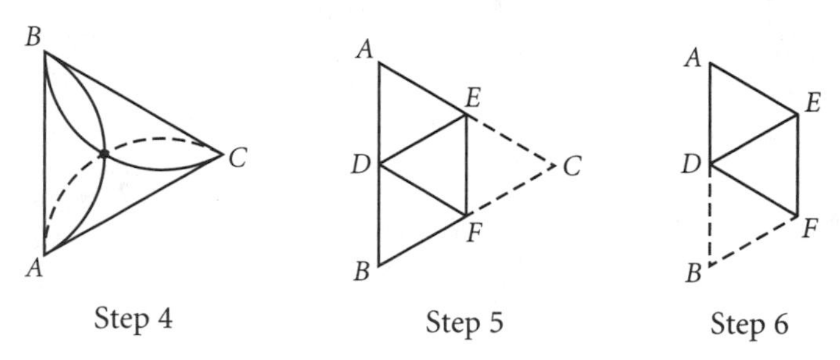

Step 4 Step 5 Step 6

(continued)

Step 7 Fold point A across to point F along segment DE. What type of triangle did you form? How do you know? How does it compare to the first triangle? Unfold and compare the areas of the two triangles.

Step 8 Open back to $\triangle ABC$. Form a three-dimensional solid by folding along $\overline{DE}$, $\overline{EF}$, and $\overline{DF}$. This solid is called a **tetrahedron.** Discuss its properties. Be as specific as possible.

Step 9 Fold point A to the midpoint of $\overline{DE}$, point B to the midpoint of $\overline{DF}$, and point C to the midpoint of $\overline{EF}$. Lay the smallest triangle flaps (points A, B, and C) flat on each other to make a solid called a **truncated tetrahedron.** Discuss its properties. Be as specific as possible.

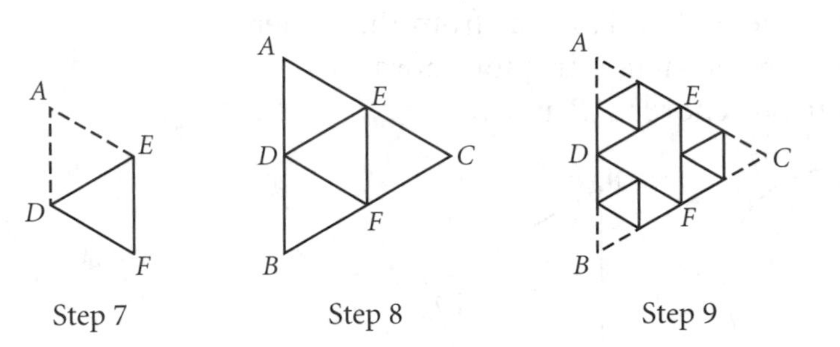

Step 10 Unfold to the original circle. Discuss the geometry concepts or geometric figures shown within the folded lines. Decorate your final product colorfully to show a particular geometry property.

PROJECT
The Art of Pi

You know that the ratio of the circumference of a circle to its diameter is a number called π. But what actually is this number? Why does it go on forever without repeating? Why is the definition so simple and the actual number so complicated? Many people throughout history have been fascinated with π. Mathematicians have used supercomputers to calculate billions of digits of π, far more than could ever be useful. Artists, poets, and composers have also created works that use π in some way. For this project, you will create your own work of art using π. You might want to research different art pieces that other people have made using π. You can follow the links at www.keymath.com/DG to get some ideas. What can you create using π? You might write a poem in which the number of letters in successive words is the value of successive digits of π. Or you might compose a piece of music using notes to represent the digits. The possibilities are endless!

PROJECT

Racetrack Geometry

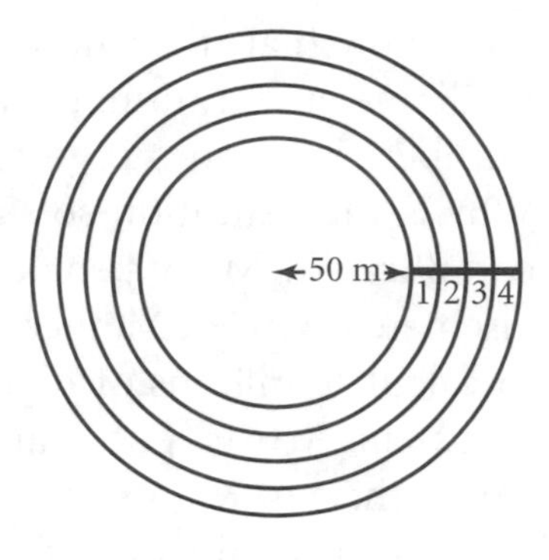

If you had to start and finish at the same line, which lane of the racetrack shown at right would you choose to run in? Sure, the inside lane. If the runners in the four lanes were to start and finish at the line shown, the runner in the inside lane would have an obvious advantage because that lane is the shortest. For a race to be fair, runners in the outside lanes must be given head starts.

Your task in this project is to design a four-lane oval track with straightaways and semicircular ends. The semicircular ends must have inner diameters of 50 meters, and the distance of one lap in the inner lane must be 800 meters. Draw starting and stopping segments in each lane so that an 800-meter race can be run in all four lanes.

What do you need to know to design such a track? You will need to determine the length of the straightaways. You will also need to determine the head start for each of the runners in the outer lanes so that each runs 800 meters to the finish line. Before you begin creating your racetrack, you will need to answer these questions.

- Does the radius of the circle play a part in determining the head start?
- Does the width of the lane play a part in determining the head start?
- Do the lengths of the straightaways play a part in determining the head start?

To answer these questions, try calculating the lengths of a few sample racetracks. For example, if the inner radius of the circular track shown above is 50 meters and each lane is 1 meter wide, you can calculate the distance each runner must travel in one lap, staying in his or her own lane.

Complete the table below. The radius in the table is the radius of the circle that defines the inside edge of each lane.

Circular Track
Inner Radius 50 m, Lane Width 1 m

Lane	Radius (m)	Circumference (m)
1	50	100π
2	51	
3		
4		

To make your race fair, look in your table to determine how much of a head start each runner in the outer lanes must have. For the circular track shown above, you can see that the runner in lane 2 must have a head start of 2π meters over the runner in lane 1; the runner in lane 3 must have a 2π-meters head start over the runner in lane 2; and so on. With these head starts, each runner will travel 100π meters.

(continued)

Is the head start always 2π meters? Investigate a few other tracks to find out. You can either complete the tables below or make up some of your own.

Circular Track
Inner Radius 65 m, Lane Width 1 m

Lane	Radius (m)	Circumference (m)
1	65	130π
2	66	
3		
4		

Circular Track
Inner Radius 65 m, Lane Width 1.5 m

Lane	Radius (m)	Circumference (m)
1	65	130π
2	66.5	
3		
4		

From these examples you may be able to answer the first two questions. Most tracks go around a playing field and have straightaways. What about the length of the straightaways? Complete the tables below to calculate the total distance for each lane in the type of track shown below.

Oval Track
Inner Radius 30 m, Straightaway 100 m, Lane Width 1 m

Lane	Radius (m)	Straightaway (m)	Total distance (m)
1	30	100	$200 + 60\pi$
2	31	100	
3	32	100	
4	33	100	

Oval Track
Inner Radius 30 m, Straightaway 200 m, Lane Width 1 m

Lane	Radius (m)	Straightaway (m)	Total distance (m)
1	30	200	$400 + 60\pi$
2	31	200	
3	32	200	
4	33	200	

From these tables you may be able to answer the third question. If you can answer all three questions, you are ready to start designing your racetrack.

Again, your task in this project is to design a four-lane oval track with straightaways and semicircular ends. The semicircular ends must have inner diameters of 50 meters, and the distance of one lap in the inner lane must be 800 meters. You determine a width for the lanes. Draw starting and stopping segments in each lane so that an 800-meter race can be run in all four lanes.

On your mark, get set, GO!

PROJECT • Parallelogram Tessellations Using the Midpoint Method

Students create a tessellation by using a point-symmetric curve on each side of a parallelogram. This project can supplement Lesson 7.7. The project could also be completed using geometry software.

Outcome

- Student completes a tessellation using the midpoint method.

PROJECT • Tessellation T-shirts

contributed by Carolyn Sessions

This process can be used to transfer a tessellation to a T-shirt or any piece of fabric. The best kind of T-shirt to use is a 50/50 cotton-polyester blend, rather than a shirt made of 100% cotton. If your students can't afford T-shirts, buy white muslin from a discount store and cut it into squares. Students can transfer their tessellations to the squares, and you can make a class quilt by using tape or fusible hem tape to hold the squares together.

Materials

- T-shirts or pieces of fabric
- fabric crayons (available at school-supply, arts-and-crafts, and some fabric stores)
- tracing paper (Avoid using patty paper for this project as it has a finish that comes off on the shirt when it is heated.)
- iron and ironing board
- large sheets of clean, white paper or blank newsprint (Sheets from an artist's sketchpad work well, but any white, nonwaxed paper should work.)

Instructions

1. Ask students to draw a design tile that will tessellate on a sheet of white paper. Then have students use a pencil to tessellate their design onto a sheet of tracing paper so that the tessellation covers the entire sheet of paper. Students should trace the outline and details for each design tile.
2. Once the tessellation has been traced onto the tracing paper, have students use fabric crayons to color the *back* side of the paper. Make sure they don't color the side on which the tessellation has been traced. Some students also like to color a single design tile for the shirt's front pocket area.
3. Set the iron to its hottest setting and place the shirt on the ironing board. Put a sheet of clean, white paper inside the shirt, between the front and the back, so that the colors don't bleed through both sides of the shirt.
4. Smooth wrinkles from the shirt. Center the tessellation on the front or back of the shirt, about 2 inches below the neckline. Make sure the colored side of the tessellation is next to the fabric.
5. Cover the tessellation with a sheet of clean, white paper. Iron on the white paper, pressing lightly and making sure to iron over all the edges. If you find that the iron is starting to scorch the fabric, use a lower temperature setting or more white paper as protective thickness.
6. To check for a complete transfer, anchor the center of the white paper and the tracing paper with one finger. Carefully lift the edges of both pieces of paper. If not all parts of the tessellation are evenly transferred, reposition it carefully and iron again over the lightly transferred areas. Again, watch for scorching.

PROJECT • Stained Glass

contributed by Sharon Grand Taylor and Carolyn Sessions

Beginning in Chapter 3, students acquire and practice construction skills requiring the use of compass and straightedge or patty paper. Constructions are used to lead to students' discovering geometric relationships and making conjectures; this project reinforces these skills and allows students to combine them with their creativity and artistic talents.

"Stained glass" is really painting on glass: a water-based simulated liquid leading is used to outline the constructed design, and then the outlined areas are colored by filling them with water-based glass paints, or "stains." The paints become clear or translucent when dry, so when the finished pieces are hung in a window, they have the appearance of stained glass.

This project is popular with students because they not only review geometry concepts and use problem-solving skills to plan and create their designs, but also connect geometry and art to create something decorative that others will see and admire. Students who might not be great successes academically can shine with this project. Many have reported that they still have their stained-glass pieces years after graduating from high school.

The stained-glass instructions can be used for a variety of student work that has been designed on paper: compass-and-straightedge designs, tessellations, geometric art designs. It is difficult to work with designs that have very small areas to color. Designs can be originals or re-creations of designs in construction books or posters. This project could also be adapted for use with Chapter 0 or Chapter 3.

It is helpful to work in a room with access to water to make cleanup easier. A science lab with faucets and tables is ideal. The finished pieces can be collected in an eye-catching display. Rather than putting the glass into the frame, affix two 1 in. strips of double-sided foam tape to the back of the glass. Press them onto the window glass. These can be removed by inserting a thin, flexible blade (such as a palette knife) between the two pieces of glass to break the foam tape.

MATERIALS

- 8-by-10 in. pieces of clear glass or 8-by-10 in. wall picture frames with regular glass. The picture frames should be wood frames that have staples on the back to keep the glass in. Do not get frames with nonglare glass!
- water-based glass stains
- small watercolor brushes
- one small bottle of liquid leading (water-based) per group of four students
- dish soap to clean brushes
- newspaper to protect surfaces from paint
- eyedroppers
- cotton swabs
- compass or large safety pin
- scissors

PROJECT

Parallelogram Tessellations Using the Midpoint Method

In Lesson 7.7, you saw several methods you can use to create tessellations by using rotations. There are various other methods you can use. In this project you'll create a tessellation by using a point-symmetric curve on each side of a parallelogram. Follow the steps below to create your tessellation.

Step 1 Construct a parallelogram grid and the midpoints of one parallelogram.

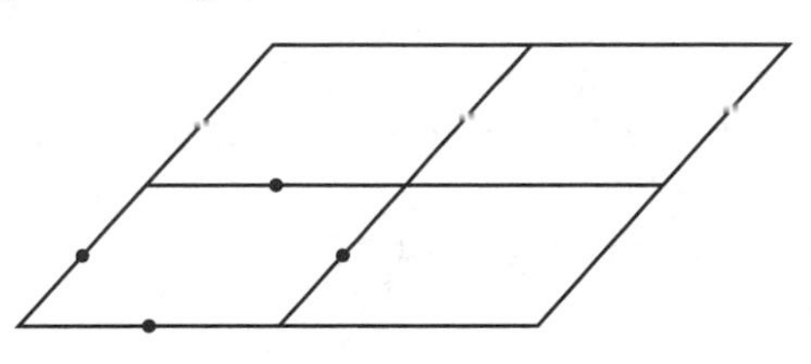

Step 2 Draw a curve on half of one side of the parallelogram and carefully rotate it about the midpoint of that side. Do the same for the other three sides.

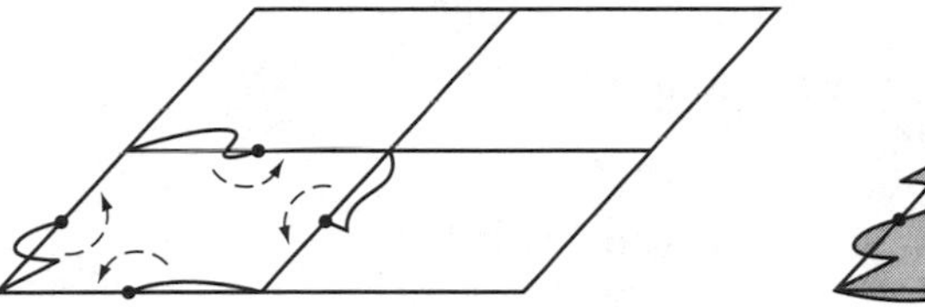

Step 3 Tessellate. Decorate your creation.

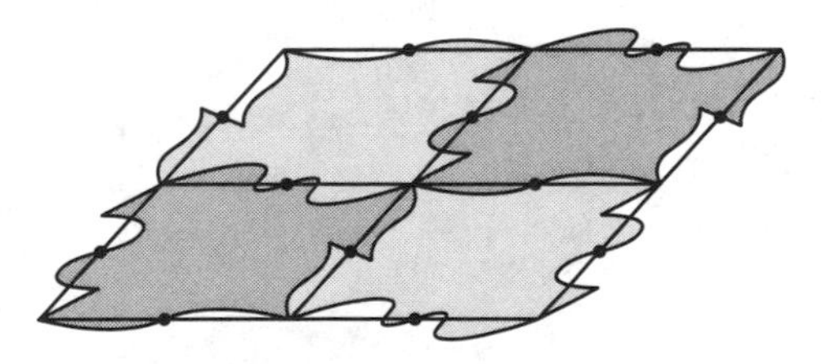

PROJECT

Tessellation T-shirts

In this project you will transfer your own tessellation to a T-shirt or a piece of fabric. (The best kind of T-shirt to use is a 50/50 cotton-polyester blend, rather than a shirt made of 100% cotton.)

Part 1: Creating Your Tessellation

Step 1 Look through Chapter 7 of your book and use one of the methods described to make your own design to tessellate. Include detail in your design.

Step 2 On tracing paper, use a pencil to carefully tessellate your design so that the tessellation covers the entire sheet of paper. Be sure to trace your design's detail, too. Leave room for the title of your tessellation and your signature (see Steps 3 and 4).

Step 3 Title your tessellation and write the title on the tracing paper where you think it will look best. Make your letters large enough to be seen from a distance.

Step 4 Sign your tessellation with your name. If you have a "famous" signature, use it. You may want to make your signature a little larger than you normally would.

Step 5 Bring your masterpiece to school so that you can transfer it to your shirt.

Part 2: Preparing Your Tessellation for Transfer

Step 1 On the back of your tracing paper, draw over the outline of your tessellation with a dark-colored fabric crayon. Then color the interior of the tessellation.

Step 2 Outline and color your title.

Step 3 Outline your signature with a dark-colored crayon.

Step 4 If you want a design on your shirt's front pocket area, use a small sheet of tracing paper to trace, outline, and color a single design tile. Remember, color the back of the traced design tile.

Step 5 Ask your teacher or an assigned "peer helper" to help you transfer your tessellation to your T-shirt.

PROJECT

Stained Glass

In this project you will transfer a geometric design onto glass and create a stained-glass window. It is difficult to work with designs that have very small areas to color. Keep this in mind when you create your design. Designs can be originals or re-creations of designs in construction books or posters. You can also look in Chapter 0 and Chapter 7 of your book for some ideas.

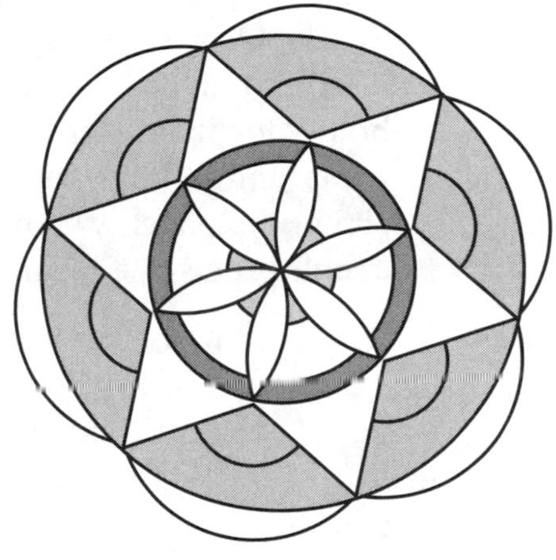

Part 1: At-Home Preparation

Step 1 You will need an 8-by-10-in. piece of regular glass. Do not get nonglare glass; the stain will not stick to nonglare glass. You can purchase an 8-by-10-in. wood picture frame that has staples on the back to keep the glass in place. This way, you will have a frame ready for your stained glass when you have finished the project.

Step 2 If you bought glass in a frame, take the glass out of the frame. Place masking tape around all the edges of the glass so that you won't cut yourself or someone else with an edge.

Step 3 Wash the glass with soap and water. Do not use glass cleaner; if you do, the paint will not stick to the glass. Dry the glass thoroughly with a paper towel or lint-free towel.

Step 4 Wrap the glass in several layers of newspaper and tape them securely around the glass.

Step 5 Put the glass (and the frame, if you bought one) into a paper bag and bring it to school. Bring some extra newspaper to cover your desk.

Step 6 Dress in old clothes on painting day. The paint may stain your clothes.

Step 7 Draw or construct a geometric design to use on your glass. The design should fit into a rectangle that is about an inch smaller than the glass in all directions. Otherwise, the frame may cover part of your design. Use watercolors or markers to select colors for your design.

Part 2: Painting the Glass

Step 1 Clear your workspace of unnecessary items and place a protective layer of newspaper on it. Place your design on the table and put the glass on top of it. Arrange the glass until the design is aligned the way that you would like it to appear on the glass when framed.

Step 2 Tape the design to the glass at the upper and lower edges so that the paper will not move while you are painting.

(continued)

Step 3 Get a bottle of liquid leading. Cut the end of the nozzle. After recapping it, shake the bottle so that the liquid leading fills the nozzle. On a sheet of newspaper, practice making a smooth line of "lead" come out of the nozzle. Squeeze with even pressure and hold your hand steady. Try resting the tip of the nozzle on the paper and holding the bottle at an angle and slowly dragging it across the paper. When you are confident with the leading, you can use the same technique to apply your design to the glass.

Step 4 Use the liquid leading to outline your design on the glass. Do not worry if all of the lines are not uniform in size. Make the lines rather thick. If there are any breaks in the lines, go back and fill them in. Use a pencil point or a compass to reopen the nozzle if it clogs up.

Step 5 If possible, let the liquid leading dry overnight. Otherwise, begin adding the paint stain to your design.

Step 6 Use an eyedropper or a watercolor brush to paint the areas between the lines of leading. You want to make as thick a coat of paint as possible without having it spill over the lines into adjacent regions. Eyedroppers are helpful in removing air bubbles; cotton swabs are useful in cleaning up spills. Make sure the stain adheres to the edges of the leading strips.

Step 7 Let the paint dry overnight. When your stained glass is completely dry, you can hang it in a window, either by using double-sided foam tape or by attaching a chain and small screws to the top of the wood frame.

PROJECT • Building a Home

contributed by Ralph Bothe

Students use the floor plan of a home to estimate the cost of various finishing jobs such as painting, carpeting, and tiling. To estimate the costs, students must contact suppliers, provide samples, and justify their choices. The project is a natural extension of some of the problems in Chapter 8.

MATERIALS

- Sample Floor Plan worksheet, *optional*

OUTCOME

- Students provide calculations, samples, and cost estimates for each of the six jobs.

PROJECT • An Occupational Speaker

contributed by Ralph Bothe

This project contains student guidelines for inviting a guest speaker to give a class presentation about how she or he uses geometry in her or his job. It is a general-interest project that can be used at any time during the course.

OUTCOMES

- Student prepares and helps guest speaker give a successful presentation.
- Student gives an evaluation survey to the class and collates results.
- Written summary is well organized and complete.

PROJECT

Building a Home

When architects and builders work with a client, they must do a careful analysis to estimate costs of renovating an existing home or building a new home or office. Based on the information in the blueprints, they calculate the areas of all the rooms, walls, and ceilings to determine the amounts and costs of materials such as paint, carpet, tile, and sheetrock. They use these calculations to estimate the amount of money their client will have to pay to complete a job. The challenge is to keep costs low enough to satisfy the client and to choose materials that look nice and will last.

For this project you will determine costs to do some of the jobs required to finish building a home. You can use the Sample Floor Plan worksheet, the floor plan for your own home, plans in an architectural magazine or book, or, if you are ambitious, your own scale floor plan of your dream home. Estimate the cost of each of the following jobs for your floor plan.

1. Carpet the three bedrooms. (Don't forget about the carpet pad.)
2. Put in hardwood floors for the living room and dining room.
3. Sheetrock the garage walls and ceiling. (Assume a 10 ft ceiling.)
4. Tile the bathroom and front entry.
5. Paint the living room and dining room. (Assume a 12 ft ceiling.)
6. Put baseboard in all rooms that are measured in the floor plan.

For each job that you are estimating, follow these steps.

Step 1 Research the product you need. Find out what sizes it comes in and how you buy it (by the yard or by the square foot, for example). List the names of at least two suppliers that sell the product. If possible, obtain literature about the product showing samples, colors, and types. Choose a particular product to use.

Step 2 Calculate the quantity needed. Show all calculations in detail and show how you arrived at the quantity needed. Explain how you would order it from the supplier.

Step 3 Obtain at least two bids for the job to be done. Get bids with and without labor. Find out if there are different levels of quality for the product and why you might want a certain level of quality.

Step 4 Select one of the bids and explain why you picked that bid.

Step 5 Write a clear itemized cost estimate for each job as if you were a contractor bidding the job for a client. You may want to include an alternate choice, with an explanation of the possible benefits of choosing one option over the other. In addition, include the sample materials, detailed calculations, and final cost with and without labor.

Sample Floor Plan

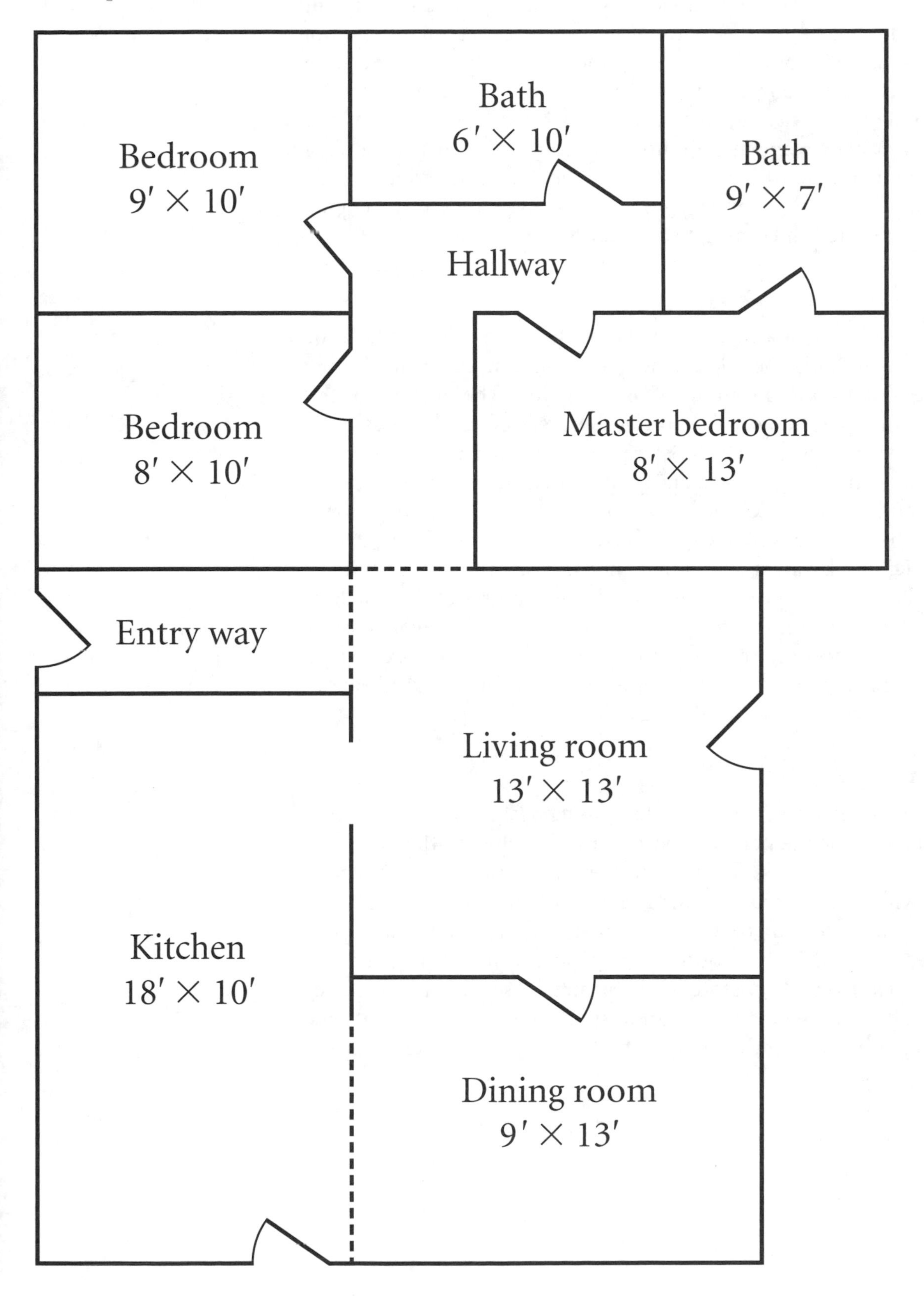

PROJECT

An Occupational Speaker

Architects, carpenters, plumbers, doctors, dentists, and artists are all people who use geometry daily. In fact, if you talk to people in different professions, you will be surprised to find out that almost any occupation requires some use of geometry. For this project you will arrange to have a guest speaker come to your geometry class and do a presentation to the class on how he or she uses geometry in his or her job. This person can be a relative, a friend, or someone else from the community whose profession uses geometry. If you need help finding someone, try contacting volunteer and youth organizations. They may be able to help you identify people who will be interested in coming to a classroom.

Interview

Once you have identified a person who wants to come to your classroom, contact the person and schedule a time when it is convenient for him or her to spend about 30 to 45 minutes talking to you. The purpose of this interview is to prepare the speaker to give an overview to the class of all the ways he or she uses geometry. Before you start your interview, you should prepare a list of questions and be prepared to take notes. Your lead question should be "How do you use geometry in your job?" If you are confused by the explanation, ask questions until you are satisfied that you understand. Remember, this person may not be used to working with students, so be patient and respectful. If this person uses any particular charts or instruments, make sure you understand how they work. Ask the person for help in creating three or more practical problems relating to part of his or her job. Write the problems on a worksheet and make sure the speaker is prepared to cover them in class.

Preparation

Talk to your geometry teacher and to the guest speaker and find a time that is convenient for the person to visit your class. Before the visit, arrange to have a guest pass for your speaker and a pass from your teacher so that you will be able to meet your guest at the main office when he or she arrives. Make sure the problems you created are included during or after the presentation. All handouts and worksheets should be given to the teacher at least one day before the presentation so that an adequate number of copies can be made. You are responsible for all handouts and notes. Arrange for any extra equipment that might be necessary at least one day in advance.

(continued)

Presentation

Meet the speaker at the door or in the main office on the day of your presentation. Introduce the speaker to the teacher and to the class. Assist the speaker with any handouts or equipment. Walk the speaker back to the door of the school when the presentation is over. After the presentation, write a thank-you note to the speaker. Call the speaker a day or so after the presentation, thanking him or her again and asking a few evaluation questions—for instance, "Did you feel the presentation was successful?" Prepare and distribute an evaluation survey for your class. Collate your survey results and share them with your teacher and your class. You might want to share them with the speaker as well.

Written Summary

Turn in a written summary of the project to your teacher. This should include a short description of the interview, the name and phone number of the speaker, the problems proposed, your own evaluation, a summary of the class evaluations, copies of all handouts, worksheets, and notes used, and anything else you find significant to the project.

PROJECT • Creating a Geometry Flip Book

Students create a flip book demonstrating an area formula or a Pythagorean Theorem dissection. This is a guided version of the project in Lesson 9.1 of the student book. Students could also complete the project using geometry software.

OUTCOMES

- Movement in the flip book is smooth.
- The figures are clearly labeled, and the explanation is complete and correct.

Extra Credit

- The student adds another animation.

PROJECT • Finding Another Proof of the Pythagorean Theorem

contributed by Masha Albrecht

Students research a Pythagorean Theorem proof they have not studied in class, then present and explain the proof. This project can supplement Lesson 9.1. You might also use it with Chapter 13.

OUTCOMES

- A proof not studied in class is presented.
- The explanation is clear and complete.
- The project includes a description of the discoverer of the proof and a list of resources.

TECHNOLOGY EXPLORATION • Graphing Circles and Tangents

Students use graphing calculators to find the equations of tangent lines. This activity can supplement Lesson 9.5.

MATERIALS

- graphing calculators

GUIDING THE ACTIVITY

Step 1: $x^2 + y^2 = 25$; $\frac{4}{3}$

Step 2: $-\frac{3}{4}$

Step 3: $y = -\frac{3}{4}x + \frac{25}{4}$

Step 4: $y = \sqrt{25 - x^2}$, $y = -\sqrt{25 - x^2}$. Make sure students understand what a friendly window is.

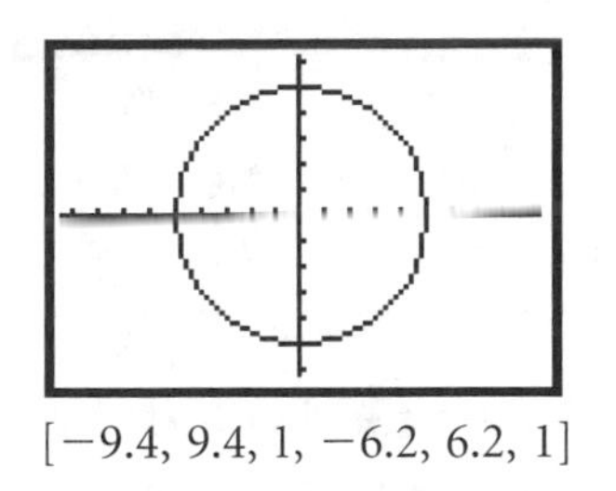

[−9.4, 9.4, 1, −6.2, 6.2, 1]

Step 5: Possible answer: Using (1.4, 4.8), the slope of the radius is $\frac{4.8}{1.4}$, or about 3.43. The perpendicular line through this point is $y = -\frac{1.4}{4.8}x + \frac{25}{4.8}$, or approximately $y = -0.29x + 5.21$. This is the tangent line at that point.

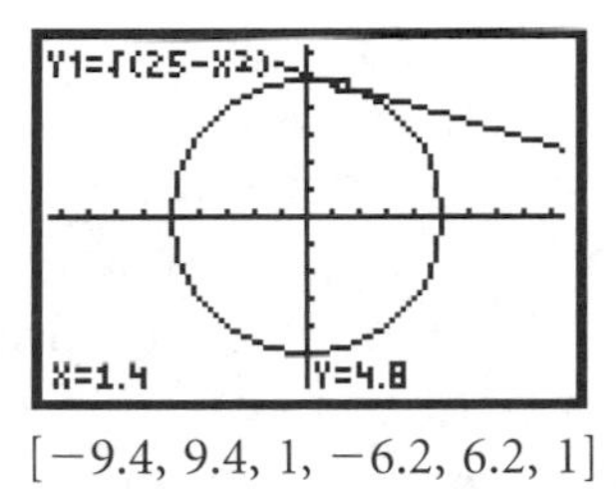

[−9.4, 9.4, 1, −6.2, 6.2, 1]

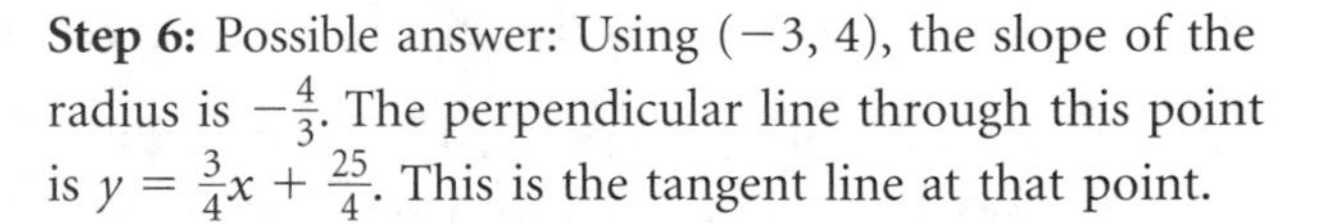

Step 6: Possible answer: Using (−3, 4), the slope of the radius is $-\frac{4}{3}$. The perpendicular line through this point is $y = \frac{3}{4}x + \frac{25}{4}$. This is the tangent line at that point.

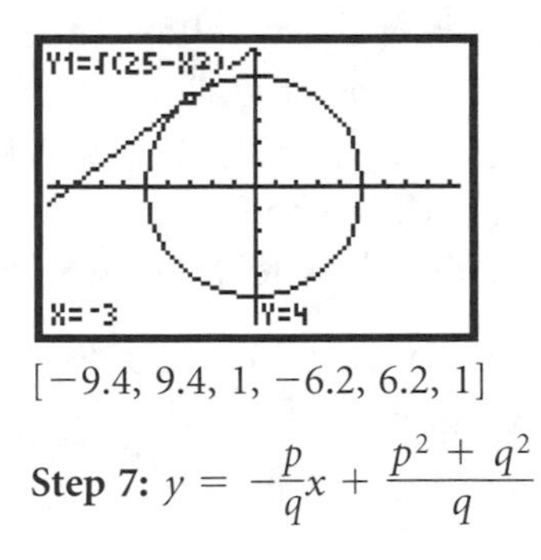

[−9.4, 9.4, 1, −6.2, 6.2, 1]

Step 7: $y = -\frac{p}{q}x + \frac{p^2 + q^2}{q}$

PROJECT

Creating a Geometry Flip Book

Have you ever played with a flip book? You can find them in novelty stores or stationery stores. The pages of a flip book contain a sequence of cartoons. Each cartoon is slightly different from the previous one. When you fan the pages quickly with your thumb and fingers, the cartoons seem to come alive and move.

The flip-book technique is basic to the creation of animation. From Walt Disney's Mickey Mouse of the 1920s to today's Bart Simpson, animators use a sequence of slowly changing cartoons to create the illusion of motion: animation. Today, animators working on computers use these same animation ideas to create special effects for movies and television.

In this project you will create a flip book demonstrating either an area formula or the Pythagorean Theorem.

Creating an Area-Formula Flip Book

In Chapter 8, you discovered that you can cut up a circle into sectors and rearrange them into a "parallelogram." Here are a few other ideas for area-formula flip books.

- Cut a triangle along a midsegment, dividing it into a smaller triangle and a trapezoid. (If the triangle is obtuse, the midsegment must be parallel to the side opposite the obtuse angle.) Drop an altitude to the midsegment from the vertex opposite the midsegment to divide the small triangle into two right triangles. Rotate each of the two right triangles about the midsegment endpoints, and match the triangles' sides with the trapezoid's legs to form a rectangle.
- Cut a trapezoid along its midsegment, dividing it into two trapezoids. Drop an altitude to the midsegment from one of the vertices of the smaller trapezoid, dividing the smaller trapezoid into a right triangle and another trapezoid. Rotate the right triangle about the midsegment endpoint that is closest to it, and rotate the trapezoid about the other midsegment endpoint. Match the triangle's side and the trapezoid's leg with the larger trapezoid's legs to form a rectangle.

You can also create your own area-formula flip book. Try converting a regular octagon into a parallelogram and then into a rectangle, or find a way to convert a rectangle into a square.

(continued)

Creating a Pythagorean Theorem Flip Book

There are many ways to show that the sum of the areas of the squares on the two legs of a right triangle is equal to the area of the square on the hypotenuse. You discovered one such dissection in the investigation in Lesson 9.1 of your book. Here are two other dissections you could use to create a Pythagorean Theorem flip book.

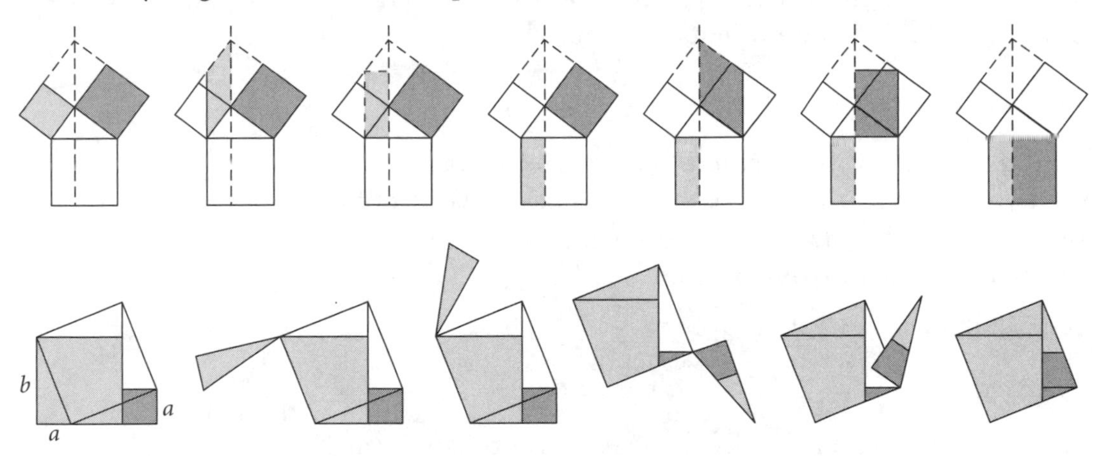

You can use any of the dissections from this book for a Pythagorean Theorem flip book. You can also research other proofs of the theorem to make a flip book of your own.

Constructing Your Flip Book

1. On graph paper, draw your nonmoving figures in the same position on each page. This will ensure that they don't appear to jump around or vibrate when you flip the pages of the flip book. Try to find a small graph-paper tablet for your flip book.
2. If you can't find a small graph-paper tablet, you can do your work on thin transparent paper. Use transparent paper to trace your nonmoving features so that in each successive drawing only your moving shapes will be drawn in a new position. Glue the transparent paper onto three-by-five cards for a firmer flip card.
3. The more steps you make in your animation, the more slowly the pieces move from page to page. This gives a smoother motion.
4. Explain what your flip book demonstrates. If you chose a Pythagorean Theorem dissection, explain how your flip book demonstrates the Pythagorean Theorem. Label each picture so that it's clear how the process works.
5. If there is a budding artist in your group, perhaps you can add other animation to your flip book. For example, draw a hand flipping the pages of a flip book on each page, thus creating a flip book of a flip book!

PROJECT

Finding Another Proof of the Pythagorean Theorem

Many people have found original proofs of the Pythagorean Theorem. Historical documents show ancient proofs of this famous theorem by mathematicians from all over the world. In more recent times, modern mathematicians, a U.S. president, and even some high-school students have found unique proofs of the theorem. For this project, you will research a proof you have not already studied and present it to your class.

Finding a Proof

Your first task is to find a proof of this theorem that you have not already studied in class. Then you must learn enough about this particular proof to write a report or make a poster. Here are some suggestions for ways to find different proofs or resources for your project.

1. Look in your school library or local library.
2. Look on the Internet.
3. Ask math teachers if they have any books you can look through. Even a different geometry textbook can be a good resource.

Explaining the Proof

You need to explain the proof in a way that makes sense to you and your classmates. This is the hardest part of the project, because many of the proofs are complicated. Here are some steps to help.

Step 1 Make sure you understand the proof yourself. If you need another person to help you make sense of it, ask a friend, group member, relative, or teacher.

Step 2 Decide what kind of presentation would make the proof simplest to understand. Would it be clearer on a poster? Would it help to make a dynamic geometry sketch using a geometry software program? Your project should be colorful. Make sure you include lots of drawings, because these can make the project easier to read and understand.

Step 3 Create a first draft of your explanation and have a friend, group member, or relative read it over. Find someone who has about as much math knowledge as you and your classmates. If he or she does not understand parts of it, improve on these parts to make them clearer.

Step 4 Include a description of the discoverer of the proof. Try to include interesting facts about the person's life. How did he or she come up with the proof?

Step 5 Include a list of the resources you used for the project. Explain where you found the proof, and include the name of anyone who helped you.

TECHNOLOGY EXPLORATION

Graphing Circles and Tangents

In this activity you'll use algebra to find the equation of a tangent line given the coordinates of a point on a circle. Then you'll use a similar process to graph tangent lines on your graphing calculator and find the equation of a tangent line.

Activity: Equations of Tangent Lines

Step 1 Consider a circle centered at the origin and a point on the circle with coordinates (3, 4). Write the equation of the circle. What is the slope of the radius drawn to that point?

Step 2 What is the slope of a tangent line through that point?

Step 3 To graph the tangent line, you need to know the slope, m, and the y-intercept, b. You just found m. To find b, solve the equation $y = mx + b$ for b, and substitute the values you already know for y, m, and x. Once you've found b, write the equation of the tangent line.

Now that you've practiced finding an equation for a tangent line, you're ready to graph circles and tangent lines on the calculator.

Step 4 Solve your equation for the circle for y to get two equations you can graph on your calculator. Graph the circle in a friendly window with a factor of 2.

Step 5 Trace around the circle until the calculator displays friendly coordinate values. Then find the slope of the radius through this point. Find the equation of the perpendicular line through this point and graph it. Describe your result.

Step 6 Trace to a different point on the circle and repeat Step 5.

Step 7 Write the equation of a line tangent to a point (p, q) on a circle.

PROJECT • Euler's Formula for Polyhedrons with Holes

Students discover Euler's Formula for Polyhedrons with Holes. This project supplements the Exploration Euler's Formula for Polyhedrons.

Outcomes

- Students build four polyhedrons of genus one, and sketch or build at least two polyhedrons of genus two.
- Students give the correct formula for polyhedrons with holes: $V - E + F = 2 - 2g$, where V is the number of vertices, E the number of edges, F the number of faces, and g the genus number of the polyhedron.
- Some cases where the formula does not apply are given—for instance, if a polyhedron has a hole in one face.

EXPLORATION • The Möbius Strip

contributed by Luis Shein

Students make and explore a Möbius strip. This is a general-interest activity that extends students' ideas about what a surface is.

Materials

- paper
- scissors
- tape

Guiding the Activity

Step 1: This is a cylinder.

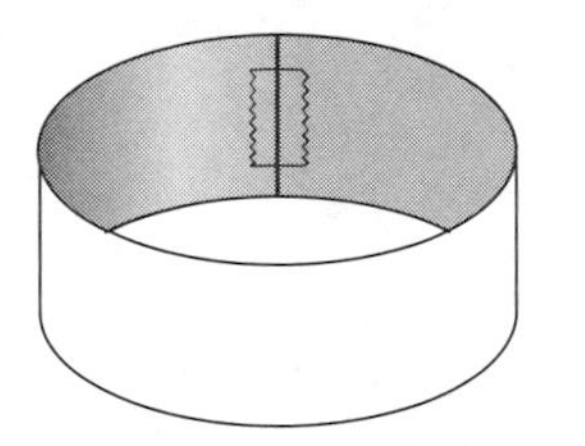

Step 2: Sample description: The Möbius strip has only one side and one edge. A path you draw on one side of the paper will cover both sides without your lifting the pencil, and if you trace along one edge, you trace both edges before returning to your starting point. The two colors meet where the strip was taped. The strip doesn't have two different sides, so you would need only one color to paint, whereas you would need two colors for the cylinder. The cylinder has two edges, rather than one, like the Möbius strip.

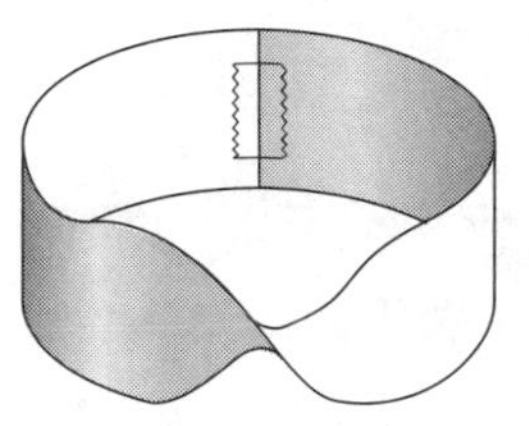

Step 3: After the fish goes once around the tank, it returns to its starting place, but it is upside down. If it rights itself, it finds that it is now facing in the opposite direction. But it *is* back where it started. You might want to have students explore other questions of handedness on a Möbius strip.

Step 4: The Möbius strip belts last longer because both "sides" get used, instead of only one.

Step 5: The actual path appears as a continuous loop on both sides of the strip, whereas the representation is two unconnected lines (one for each side of the paper strip). Opinions about the representation will vary.

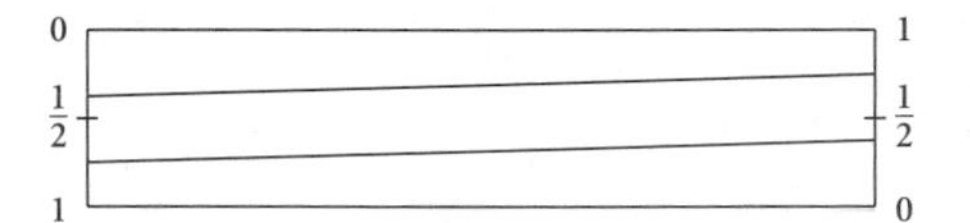

Make sure students understand that the one side of the flat representation represents the "one side" of the Möbius strip. And, for that matter, the "one side" of the twisted paper strip is not the double-long strip made of both (real) sides of the paper, but is the "inside" of the paper, so that you only have to go around it once to get back to where you started.

Step 6: Cutting the cylindrical band produces two bands, each half the width of the original band.

Step 7: Cutting the Möbius strip produces a cylindrical band half the width of the original strip with two twists (four half-twists).

Step 8: Cutting the cut Möbius strip produces two interlocking bands with two twists (four half-twists) each.

Extensions

Have students search for Möbius strips in use. They might ask a mechanic or consult a car repair manual or look for games and simulations that use Möbius strips for their surfaces. Students could also design a game using a Möbius strip themselves or research the Klein bottle.

EXPLORATION • Archimedes' Principle

Students model a boat in an ocean to explore buoyancy. This activity extends Lesson 10.5.

Materials

- empty half-gallon milk carton with the top cut off, or open plastic container
- tub, large plastic container, or sink filled with water
- rice, unpopped popcorn, or dried beans
- balance

Questions

1. The carton will float. The water will rise about 11.6 cm.
2. The box will float. The box weighs $(18 \cdot 32 \cdot 0.5 - 4 \cdot 4^2 \cdot 0.5)\ \text{cm}^3 \cdot 2.81\ \text{g/cm}^3 = 719.36$ g. Let h be the height of the water displaced: $(32 - 2 \cdot 4)(18 - 2 \cdot 4)h = 719.36$, so $h \approx 3.0$ cm. The box is 4.0 cm tall and displaces 3.0 cm of water, so it will float.
3. Norah probably took on more of a task than she should have. Each cannonball weighs about 24.2 kg above the water surface and about 22.0 kg in the water.

PROJECT • Packing Efficiency and Displacement

contributed by Carolyn Sessions

Students use displacement to measure the volume of two household items, such as a light bulb and a tube of toothpaste, then measure the volume of their packaging containers to calculate their packing efficiency. This project can supplement Lesson 10.5.

Outcome

- Student submits poster showing volume of two household items and their packaging containers, and determines which packaging is more efficient.

PROJECT • Geometry in Sculpture

Students make three-dimensional geometric art and explain the geometry in their art. This project can supplement Chapter 10.

Outcomes

- The sculpture uses geometry.
- The report is clear and complete.

PROJECT

Euler's Formula for Polyhedrons with Holes

In the Exploration Euler's Formula for Polyhedrons, you discovered a formula that relates the numbers of faces, edges, and vertices in a polyhedron with no holes. But what about polyhedrons with holes? For example, the polyhedron shown at right has one hole.

Count the vertices, edges, and faces of the polyhedron. Do the numbers work in Euler's Formula? Mathematicians call polyhedrons with no holes **polyhedrons of genus zero.** Polyhedrons with one hole are called **polyhedrons of genus one,** and so on. Follow the steps below to discover a formula that will work for polyhedrons of all genuses. You will need toothpicks and modeling clay, gumdrops, or dried peas.

Step 1 Build four different polyhedrons of genus one.

Step 2 Complete the table below for each stick polyhedron you built. Count the number of vertices (V), edges (E), and faces (F) of each polyhedron. Recall that $V - E + F = 2$ for polyhedrons of genus zero.

Polyhedron	Vertices (V)	Faces (F)	Edges (E)	$V - E + F$

Step 3 Find a formula for genus one polyhedrons. Then find a rule that works for both genus zero and genus one polyhedrons. (Try using g for genus number.)

Step 4 Sketch or build several stick polyhedrons of genus two, and find a rule that works for genus zero, genus one, and genus two. Do you think your rule would work for all genuses? What are some features a shape might have that would cause your rule to fail?

EXPLORATION
The Möbius Strip

You've seen and used many surfaces both inside and outside your geometry class. Some surfaces are flat like a plane, some are curved like a sphere, some have edges like a polygon, and some go on forever. In this activity you will create a surface that has some surprising features. This surface is called a Möbius strip. You will need paper, scissors, and tape.

Activity: The Surface with a Twist

Making the Möbius Strip

Step 1 Cut out two strips of paper that are each about 1 by 12 in. If you can, use paper that is a different color on each side. If you use white paper, color or shade one side of each strip.

Draw arrows on one strip like this:

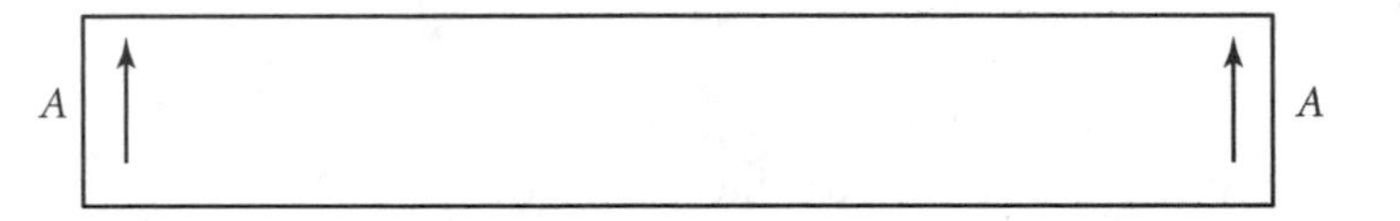

Tape together the ends of the first strip so that both arrows point in the same direction. Briefly describe the surface you have created. Include a sketch of the surface with your description.

Step 2 Draw arrows on the second strip like this:

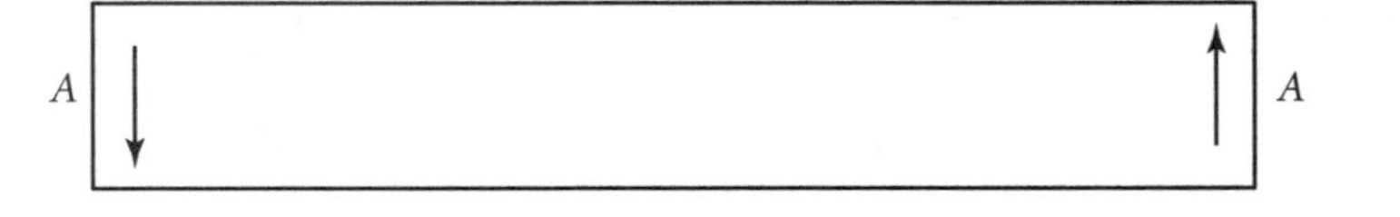

Tape together the ends of the second strip so that both arrows point in the same direction. You have just created a Möbius strip.

a. Take your pencil and draw a path along one side of the Möbius strip. Don't lift your pencil until you return to your starting place. Now trace your finger along the edge of the Möbius strip. Don't lift your finger until you return to your starting place.

b. Your original strip of paper was colored differently on different sides. Inspect the coloring of your Möbius strip. Imagine that you are painting surfaces, with the rule that each side of a surface must be a different color. How many colors will you need for the first surface you made? How many will you need for the Möbius strip?

c. Now describe the Möbius strip you have created. Include a sketch of the surface (colors would be helpful), descriptions of your results from parts a and b, and further comparisons of the Möbius strip with the simpler surface you created first.

(continued)

Step 3 Your paper Möbius strip actually has two sides (because the paper has some thickness), but a Möbius strip really has only one side. Imagine a fish tank in the shape of a Möbius strip. The tank is so thin that only a thin fish like an angelfish can swim in it. (It can't turn around, and therefore can only swim in one direction.) Now imagine that the fish swims once around the tank. Explain where the fish ends up. Is it still right-side up? Is it still facing the same direction?

Step 4 Belts that drive motors and fans are often shaped like Möbius strips. Explain why these belts last approximately twice as long as cylindrically shaped belts.

Open up your Möbius strip to turn it back into a two-sided strip. Inspect the path you drew with your pencil. The path was closed because your pencil returned to the same place where it started. Notice that the penciled path appears on both sides of the strip.

Most surfaces we read from, write on, and draw on are flat, so it's helpful to be able to represent the Möbius strip on a flat surface. One way to do this is to use a flat rectangle. Pretend the left-hand side of the rectangle is actually connected to the right-hand side, and that the rectangle has a twist.

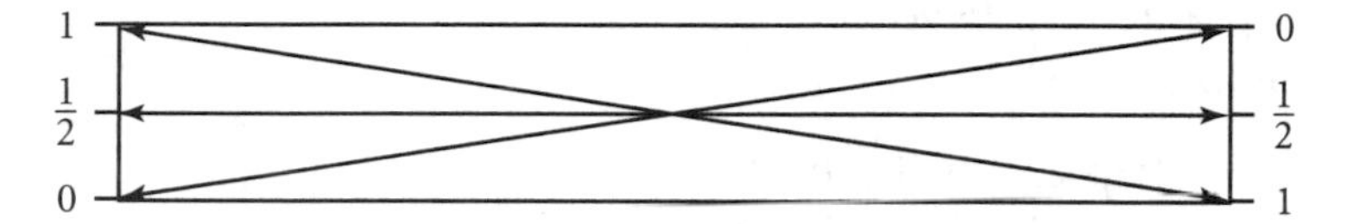

Imagine a number line from 0 to 1 on either side of the strip, like those shown above. In order to represent the twisted connection, the number lines have opposite orientations. Point 0 on one corner maps to point 0 on the opposite corner. Likewise, 1 maps to 1, $\frac{1}{2}$ maps to $\frac{1}{2}$, $\frac{1}{10}$ maps to $\frac{1}{10}$, and so on. In this representation, you do not need to use the back of the rectangle. Also, the actual length and width of the rectangle are not important.

Any path you can draw on the rectangle that starts and ends at the same number actually forms a closed path. For example, in the diagram, the path that starts at $\frac{1}{10}$ and ends at $\frac{1}{10}$ is a closed path.

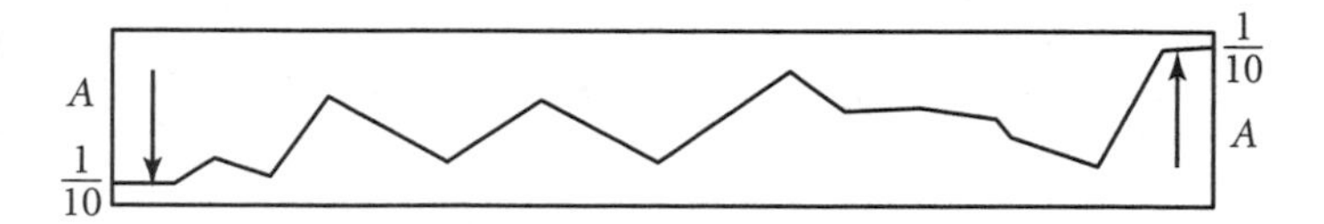

Step 5 Use this method to draw a representation of the penciled path from your Möbius strip on a flat sheet of paper. Make sure your path would be connected if it were actually on a Möbius strip. Explain how the actual path on your flattened Möbius strip appears different from the one in your representation. Describe any advantages and disadvantages you see in this rectangular representation of a Möbius strip.

(continued)

Cutting the Möbius Strip

Step 6 The diagram below shows a flat representation of the simpler cylindrical surface you made first (the one without the twist).

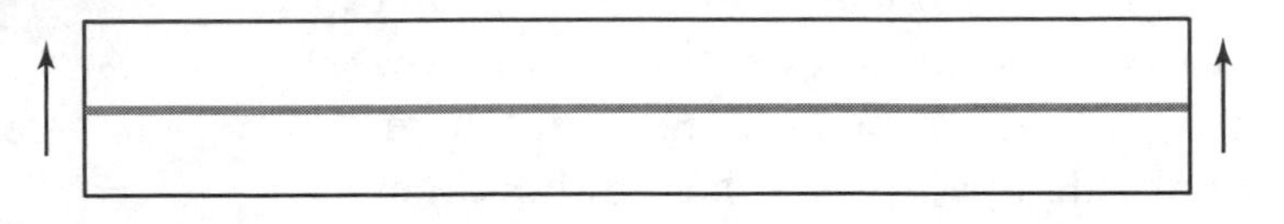

a. Imagine cutting a path on the cylindrical band from the midpoint of one side to the midpoint of the second side. Predict what will happen when you cut your cylindrical band along this line.

b. Now go ahead and cut your cylindrical band right down the middle. Describe your results. Was your prediction correct?

Step 7 This diagram shows the Möbius strip with a cut down the middle.

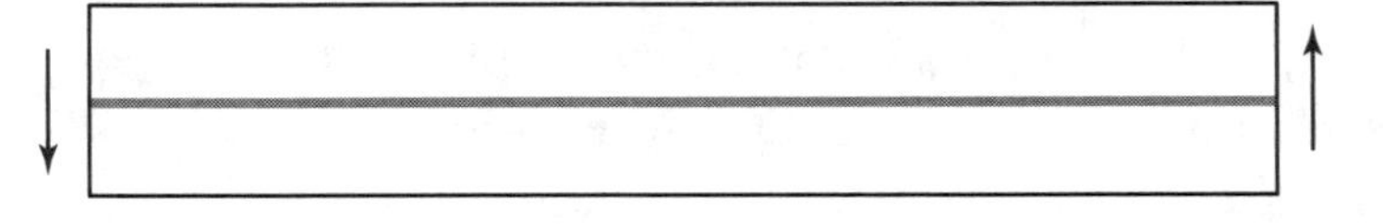

a. What do you think will happen to your Möbius strip if you cut it along a path that joins the midpoints of both sides? Think hard before you make this prediction. If you predict this one right, you have excellent visualization skills!

b. Now go ahead and cut your Möbius strip down the middle. (You may need to tape it back together again first.) Describe your results. Is this what you expected?

Step 8 For a truly unusual experience, cut your already-cut Möbius strip again. Try to predict the results before you actually do this. Record your results and any surprises you have.

EXPLORATION

Archimedes' Principle

In Lesson 10.5, you learned about displacement and density. An important property related to density and displacement is *buoyancy*. Have you ever been swimming with a friend and noticed how much easier it was to lift or pick up your friend in the water than out of it? Or have you ever played in the ocean or a lake and discovered that you could pick up a large sunken log or boulder and raise it to the surface but couldn't lift it out of the water?

The apparent loss of weight of an object submerged in liquid is called **buoyancy.** The weight lost by an object in water is equal to the weight of the water it displaces. This is called **Archimedes' principle.**

Archimedes' principle states that an object submerged in a liquid is buoyed up by a force equal to the weight of the liquid displaced. For example, if a stone weighing 250 g in air is submerged in water, where it weighs 200 g, then the water it displaces weighs the missing 50 g.

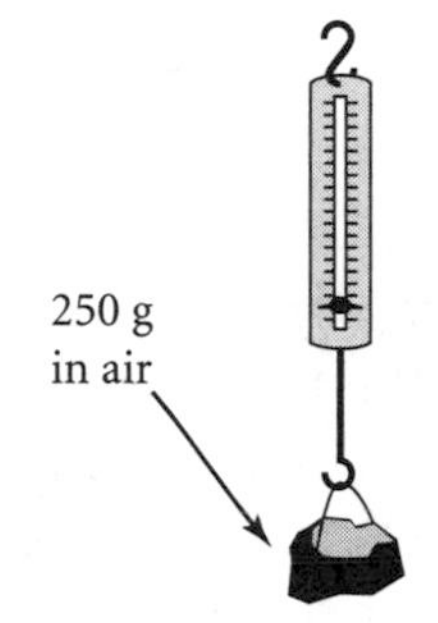

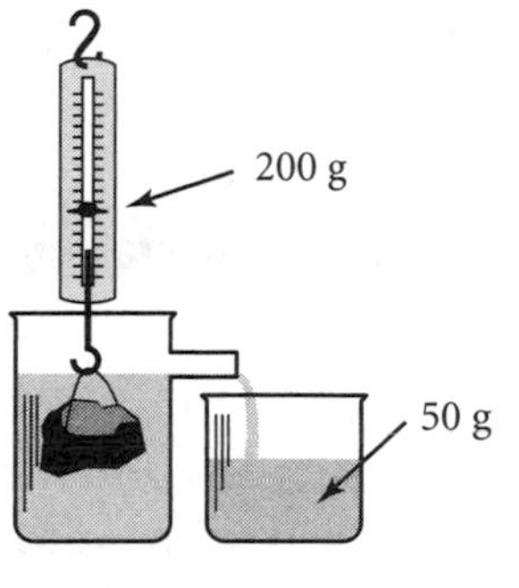

Here are some other examples: If you put a block of wood into water, it sinks until it has displaced its own weight of water, then it stops sinking. A boat floats when its weight is less than the weight of the volume of water it displaces. Submarines use Archimedes' principle. When its ballast tanks are filled with air, the submarine's weight is less than the weight of the water it displaces, and it rises to the surface. When the air is emptied from the tanks and sea water is allowed into the tanks, the submarine's weight is greater than the weight of the water it now displaces, and it sinks. To raise the submarine, compressed air is used to force the water back out of the tanks. Now the submarine once again weighs less than the water it displaces, and it floats. The same principle also works for hot-air balloons.

Example

A flat-bottomed boat (in the shape of a rectangular prism) is 4 ft wide by 6 ft high by 3 ft deep. The boat and its three potential passengers weigh 750 lb. The density of water is 63 lb/ft^3. Will the boat be able to hold all three passengers? How high up the sides of the boat will the water level reach?

(continued)

Solution

Let h be the height the water will rise. Use the fact that the weight of the water displaced equals the weight of the boat to write an equation and solve for h.

$$\text{weight of water displaced} = \text{weight of object}$$

$$\text{volume displaced} \cdot \text{density} = 750$$

$$(4)(6)h \cdot 63 = 750$$

$$h = \frac{750}{(4)(6)(63)} \approx 0.5$$

The water will rise only about 0.5 ft. Therefore, the boat can easily hold all three passengers. Could it hold three more people if they each weighed 180 lb?

Activity: Buoyancy Experiment

In this activity you'll model the buoyancy of a boat in the ocean. You will need an empty half-gallon milk carton with the top cut off or an open plastic container (for your boat); a tub, large plastic container, or sink filled with water (for your ocean); rice, unpopped popcorn, or dried beans (for your boat's cargo); and a balance to measure the weight of the boat and its cargo.

Step 1 Calculate the area of the base of your boat in square centimeters.

Step 2 Add cargo to your boat and weigh the boat with its cargo. If you are using a milk carton, start with about 500 g of cargo.

Step 3 Calculate how high the water will rise on the sides of your boat. Remember, the boat will displace an equal weight of the water. Recall that 1 cubic centimeter of water weighs 1 g.

Step 4 Place your boat with its cargo in your ocean. Measure the depth that the boat sinks into the water. Calculate the volume of the boat beneath the water surface in cubic centimeters. This volume is also the weight in grams of the displaced water. It should equal the weight of the boat plus cargo. How close were your calculations to your measurements?

Step 5 Calculate the maximum cargo weight your boat can carry without sinking. Then test your calculations. Compete with other groups to see which boat can carry the most cargo without sinking.

(continued)

Questions

1. Beans are poured into the empty rectangular carton below. The carton filled with beans weighs 1093 g. If the carton of beans is placed in water, will it float? If it floats, how high on the sides of the carton will the water level rise?

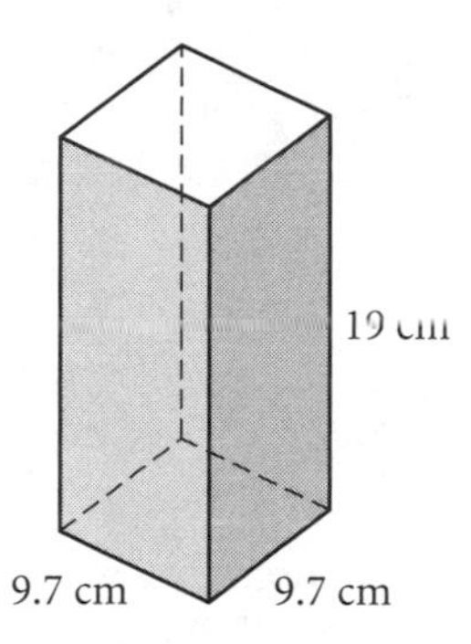

2. The sheet of aluminum at right is formed into an open rectangular box by cutting a square from each corner. When the four sides are folded up and sealed, will the box float in water? Explain. (The density of aluminum is 2.81 g/cm^3.)

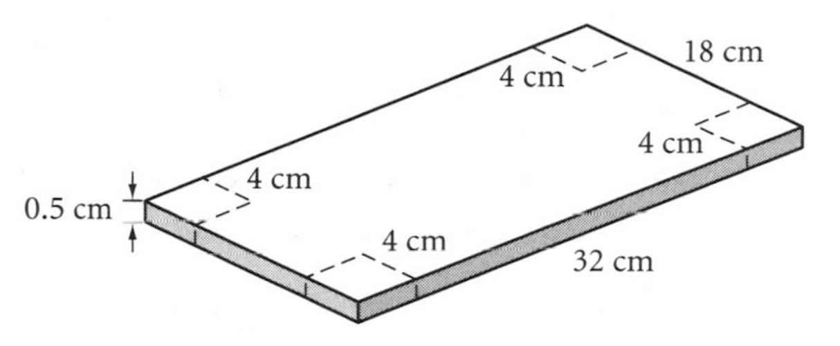

3. To impress her crew and to help maintain discipline, pirate Norah S. Grande occasionally performs feats of strength and bravery. Today, Norah's ship is anchored in the shallows off the coast, and a fishing net containing four cannonballs has been dropped over the side of the ship. Each cannonball is a solid lead sphere with diameter 16 cm. Norah announces that she will dive into the water, pick up all four balls at once, and lift them over her head. Is Norah taking on more of a task than she should? How much does each cannonball weigh when it is out of the water? How much does a cannonball weigh when it is in the water? (The density of lead is 11.30 g/cm^3, and the density of salt water is 1.03 g/cm^3.)

PROJECT

Packing Efficiency and Displacement

In this project you will use displacement to determine how efficient the packaging is for ordinary household items. You will need two different household items, such as a tube of toothpaste and a light bulb, and the boxes they came in. Make sure the items can be submerged in water! You will also need a graduated cylinder or container large enough to measure the volume of the items you chose. Make sure you can easily determine the volume of the water in the container. A rectangular prism, such as a cardboard milk carton, is usually best.

Create a single poster that answers each of the following questions. All measurements should be made to the nearest tenth of a centimeter.

1. Use displacement to find the volume of each item. Save the boxes that the items came in. Show your calculations for 1a–c.

   a. Sketch a picture of the container and the water level before the item is submerged. Label all dimensions.

   b. Sketch a picture of the container and the water level after the item is submerged. Again, label all dimensions.

   c. Determine the volume of the item.

   d. Attach the items to your poster.

2. Find the volumes of the boxes the items came in. Show your work. Attach the boxes to your poster. Label the dimensions of each box.

3. Determine the volume, in cubic centimeters, of the air space that was inside the box when the item was packaged.

4. Calculate the percentage of air space in the box by using the formula

$$\%\text{ air space} = \frac{\text{volume of air}}{\text{volume of packaging container}}$$

5. Determine which packaging container is more efficient. Explain your reasoning.

PROJECT

Geometry in Sculpture

You've seen how geometry is used in art in two dimensions. Geometric solids are also often used in art. For this project, you will create a geometric sculpture. A sculpture in this case means any three-dimensional artwork. Look through your book for some ideas. The origami boxes on page 4, the outlines of stacked cubes by Sol LeWitt on page 472, the *Verblifa tin* by M. C. Escher on page 519, and the *temari* balls on page 746 are all examples of geometric sculpture. You can use simple materials like clay, cardboard, styrofoam, and wire to make cubes, cylinders, or other shapes. You can use thread or other materials to make patterns on your shapes. You might learn how to make a Möbius strip and incorporate one in your sculpture. Or you might choose to carve a simple geometric design and its complement into two surfaces you can use to make pressed paper sculptures.

Write a short report about your sculpture, describing the geometry of your sculpture, including any symmetry it has and any geometry tools you used to create it.

CHAPTER 11 TEACHER'S NOTES

PROJECT • Similarity in Grow Creatures

contributed by Carolyn Sessions

Students use their knowledge of similar figures to determine whether two three-dimensional figures are similar. This project can supplement Lessons 11.5 and 11.6, or the Exploration Why Elephants Have Big Ears.

Grow creatures are toys that expand proportionally when placed in water (not the ones that start as pill shapes). You might introduce the project by finding a grow creature whose package indicates a given percent of growth. Ask students what they think the growth percent means. Discuss whether students think the grown creature will look like the original. Review the characteristics of similar three-dimensional figures if necessary.

Students may need some help in finding a way to measure the heights of their creatures. Suggest using an index card and placing it so that it lies parallel to another flat surface (such as a desktop) and measuring the distance between the two planes. *Note:* Tape measures work well for length and width, but plastic or wooden rulers work best for measuring height. You may want to have students weigh their creatures in class so they can use a triple beam balance.

You can find a variety of grow creatures in toy stores, or online by searching the Web for "Grow Sealife," "Grow Lizards," or "Growing Creatures."

OUTCOMES

- Student's report is clear and complete.
- Student finds that length, width, and height change linearly, whereas volume and weight change cubically.

EXPLORATION • The Golden Ratio

Students explore various properties of the golden ratio, golden rectangle, and golden spiral. This activity can supplement Chapter 11. Exercise 18 in Lesson 11.3 explores one construction for the golden cut, and Exercise 27 in Lesson 11.6 introduces the golden rectangle and golden spiral, but no prior knowledge of the golden ratio is assumed in this activity. You might use the activity with the Project In Search of the Perfect Rectangle in Lesson 11.5, in which students explore the aesthetic appeal of the ratio.

GUIDING THE ACTIVITY

Step 1: Sample description: A golden rectangle has length and width such that the ratio of the width to the length is equal to the ratio of the length to the width plus the length.

Step 2: $\phi \approx 1.618033989$

a. $\frac{1}{\phi} = \frac{\phi}{1+\phi}$, so $\phi^2 = \phi + 1$

b. $\phi^2 - \phi - 1 = 0$, so $1 = \phi^2 - \phi = \phi(\phi - 1)$, and $\frac{1}{\phi} = \phi - 1$

c. Possible answer: $\phi^3 - 3\phi = 1 - \phi$

$$\begin{aligned}\phi^3 - 3\phi &= \phi(\phi^2 - 3)\\ &= \phi((\phi + 1) - 3)\\ &= \phi(\phi - 2)\\ &= \phi^2 - 2\phi\\ &= (\phi + 1) - 2\phi\\ &= 1 - \phi\end{aligned}$$

Students might note that this quantity also equals $-\frac{1}{\phi}$ by part b.

Step 3: a. 377, 610, 987

b. 1.619048 **c.** 1.617978

d. 1.618026 **e.** 1.618037

f. 1.618033 **g.** 1.618034

h. The ratio gets closer and closer to the golden ratio.

Step 5: $ME = \sqrt{x^2 + (2x)^2} = x\sqrt{5}$

$GL = x + ME = x + x\sqrt{5} = x(1 + \sqrt{5})$

$\frac{GL}{LD} = \frac{x(1 + \sqrt{5})}{2x} = \frac{1 + \sqrt{5}}{2}$

QUESTIONS

2. The radii are $\frac{\sqrt{2}}{2}$ the radii in Step 7.

3. Sample construction:

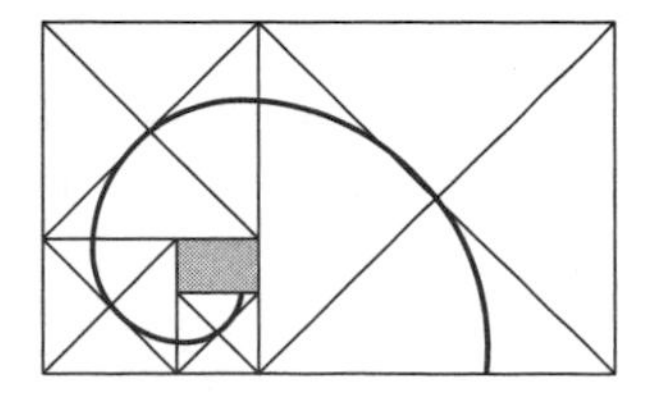

4. Sample method: Use the construction from Step 4 to find the two lengths. Construct a segment using the shorter length, then construct two intersecting arcs using the longer length.

PROJECT

Similarity in Grow Creatures

The packaging for grow creatures sometimes indicates that the creature will grow 600%. To what does this refer? Is a grown creature similar to the original? In this project you'll explore what relationships exist between the two creatures. You will need a grow creature, plain paper, a large sealable plastic bag or flat growing pan, water, tape measures or rulers, a triple-beam balance, and graduated cylinders or other containers for finding volume by displacement.

Step 1 What kind of creature do you have? Give it a name and trace the outline of your creature on a sheet of unlined paper.

Step 2 Measure the length, width, and height of your creature in centimeters. Record these measurements in the table. Mark the places where the measurements were made on your outline.

	Length (cm)	Width (cm)	Height (cm)	Volume (cm^3)	Weight (g)
Original					
Grown					

Step 3 Use a triple-beam balance to measure the weight of your creature in grams. Record the data in the table.

Step 4 Use displacement to measure the volume of your creature in cubic centimeters. Record the data in the table.

Step 5 Place your creature in your growing pan (or sealable plastic bag) and cover it with water. The water level should be about 1 cm higher than the creature. Now it's time to let your creature grow. . . .

Step 6 After the growing process, measure the dimensions of the creature in the same locations where measurements were taken earlier. Record your measurements in the table above.

Step 7 Weigh your creature. Make a drawing, label the dimensions needed to find the volume, and calculate the grown creature's volume. Did your creature grow proportionally? Justify your answer.

Step 8 Lightly squeeze some water from your creature. Place the creature in a place where it won't be disturbed to dry. Each day, record measurements until your creature has reached its original size (or very close to it). Add more rows to the table as needed.

Day	Length (cm)	Width (cm)	Height (cm)
1			

Step 9 Write a summary of the project. Did the project help you understand proportional growth and volume? Explain. Describe any patterns in the drying process.

EXPLORATION

The Golden Ratio

The **golden ratio** is a number, usually represented with the Greek letter phi (ϕ), that satisfies the special proportion

$$\frac{1}{\phi} = \frac{\phi}{1 + \phi}$$

The golden ratio is often represented geometrically by a golden rectangle. A **golden rectangle** is a rectangle whose length and width satisfy the proportion

$$\frac{w}{l} = \frac{l}{w + l}$$

w

l

A golden rectangle whose width is 1 unit has a length of ϕ units. In other words, the ratio of the length of a golden rectangle to its width is ϕ. You can use algebra to show that ϕ is equal to $\frac{1 + \sqrt{5}}{2}$.

Mathematicians have long been fascinated with the golden ratio. The ratio was of particular interest to mathematicians of ancient Greece. Some researchers believe Greek artists and architects found the golden rectangle to be the most pleasing rectangular shape. These researchers claim that golden rectangles were used many times in the design of the most famous Greek temple, the Parthenon. Look on page 630 of your book to see a photograph of the Parthenon. Whether or not Greek artists and architects actually used the golden ratio is debated, but there is no dispute that Greek mathematicians took great interest in the mathematical properties of the ratio. You'll explore some of these properties in the activity.

Activity: Properties of the Golden Ratio

Step 1 Explain the definition of the golden rectangle in your own words.

Step 2 Use your calculator to determine an approximate value for ϕ.

a. Calculate ϕ^2. How is it related to ϕ? Start with the definition of ϕ and use algebra to demonstrate the relationship.

b. Calculate $\frac{1}{\phi}$. How is it related to ϕ? Explain.

c. Calculate $\phi^3 - 3\phi$. What is this difference? Use algebra to support your discovery.

Step 3 The golden ratio can also be approximated with the help of the sequence below, called the **Fibonacci sequence.**

1, 1, 2, 3, 5, 8, 13, 21, 34, 55, 89, 144, 233, . . .

The numbers in the Fibonacci sequence occur in many branches of mathematics as well as in nature and in art.

(continued)

a. Determine the next three numbers in the sequence.

In parts b–f, use a calculator to find the ratio of each pair of consecutive Fibonacci numbers. Round your answers to six decimal places.

b. $\frac{34}{21}$ **c.** $\frac{144}{89}$ **d.** $\frac{377}{233}$ **e.** $\frac{610}{377}$ **f.** $\frac{987}{610}$ **g.** $\frac{1597}{987}$

h. What do you notice about these ratios as the Fibonacci numbers grow larger and larger?

Step 4 You can construct a golden rectangle from a square. Follow the steps to construct as large a golden rectangle as possible on an 8.5-by-11-inch sheet of paper.

a. Construct a square. Label it *GOEN*. Extend $\overleftrightarrow{GO}$ and $\overrightarrow{NE}$.

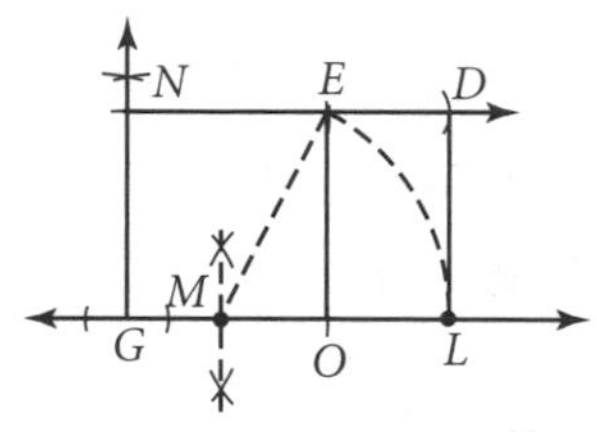

b. Bisect $\overline{GO}$. Label the midpoint *M*. With *ME* as your radius and point *M* as center, construct an arc intersecting $\overleftrightarrow{GO}$ at point *L*.

c. Construct rectangle *OLDE*. Rectangle *GLDN* is a golden rectangle.

Step 5 Use the diagram at right and the Pythagorean Theorem to show that $\frac{GL}{LD} = \frac{1+\sqrt{5}}{2}$, and thus the rectangle you constructed in Step 4 is a golden rectangle. (*Hint:* $GL = x + ME$.)

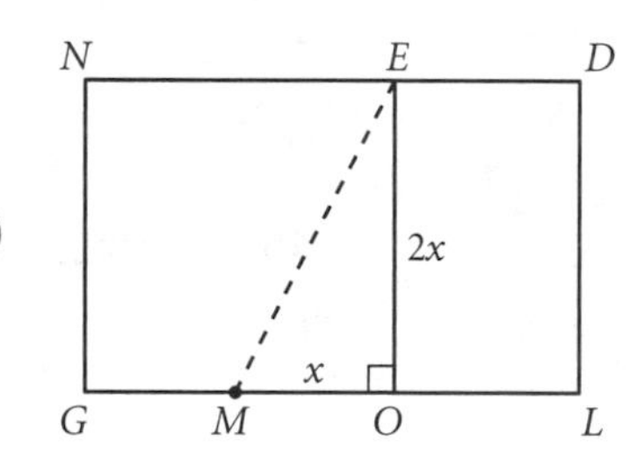

Did you notice that $\overline{OE}$ divides the golden rectangle into a square and another rectangle? Does the small rectangle appear to be similar to the original? In fact, the smaller rectangle is also a golden rectangle.

Step 6 In Step 4, you constructed a golden rectangle from a square. The small rectangle on the right side of the square was also a golden rectangle. Using the rectangle you constructed in Step 4, divide the small rectangle into a square and an even smaller golden rectangle, as shown at right. Divide this smallest golden rectangle into a square and another golden rectangle.

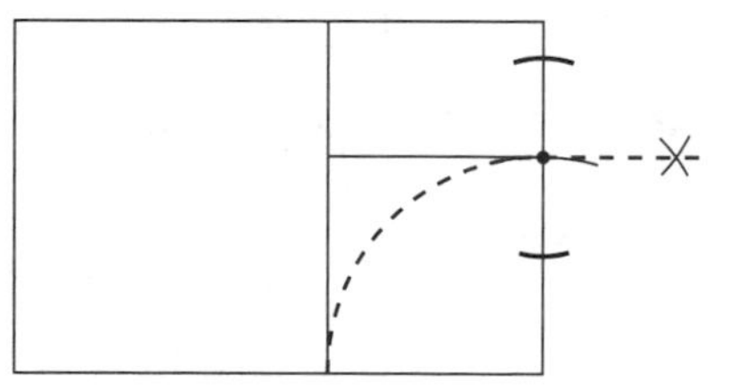

Step 7 Construct a 90° arc in each square as shown. This curve is called the **golden spiral.** It is an approximation of a curve often found in nature, which has many names, due in part to its many different but related properties. Philosopher René Descartes (1596–1650) called it the **equiangular spiral,** and astronomer Edmond Halley (1656–1742) called it the **proportional spiral.** Mathematician Jacob Bernoulli (1654–1705) named it the **logarithmic spiral** and asked that it be engraved on his tombstone.

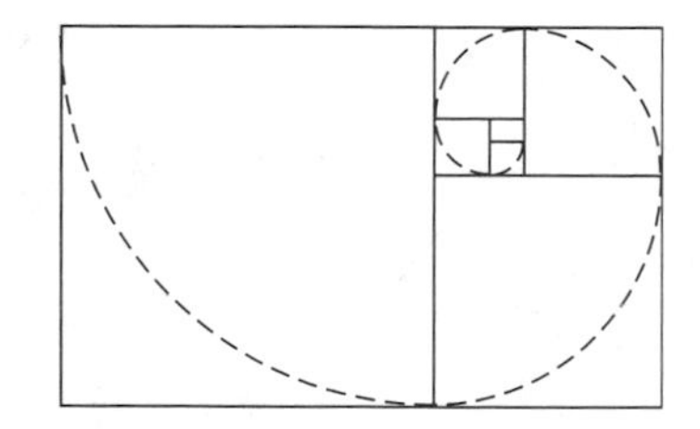

(continued)

Questions

1. The process you used in Step 6 can also be done in reverse. Construct as small a golden rectangle as possible. Use your construction tools to add a square below it, as shown in the figure in Step 6, using the long side of the rectangle as the side of the square. The square and the rectangle combine to form a larger golden rectangle. Then add another square to the left of the new rectangle, creating another larger golden rectangle. Repeat this process two times.

2. Use your golden rectangles from Question 1 to construct the approximation of the logarithmic spiral shown below. How are the radii of these arcs related to the radii of the arcs in Step 7?

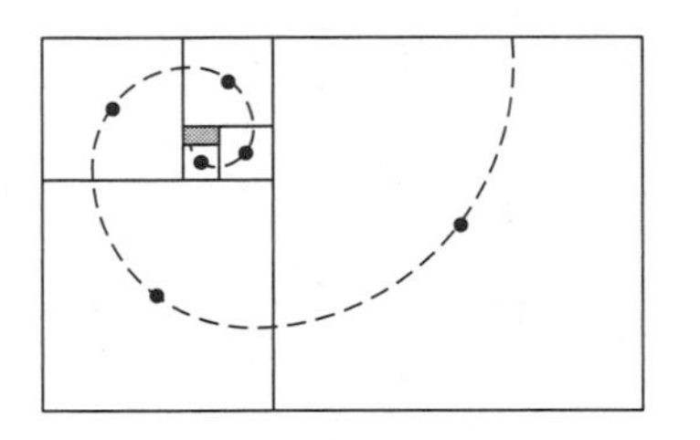

3. In his book *On Growth and Form* (1917), Scottish biologist Sir D'Arcy Thompson (1860–1948) calls the type of growth exhibited by the shell of the nautilus **gnomonic growth.** As an animal grows by gnomonic growth, the size of the animal's shell increases, but the shell's shape remains unchanged. The nautilus shell grows longer and wider at one end to make room for the growing animal within. Each new section increases in size so that the overall shape remains similar. The repeating process of adding a new square to a golden rectangle to get a new, larger golden rectangle is analogous to the nautilus adding a new chamber to its shell to accommodate its growth.

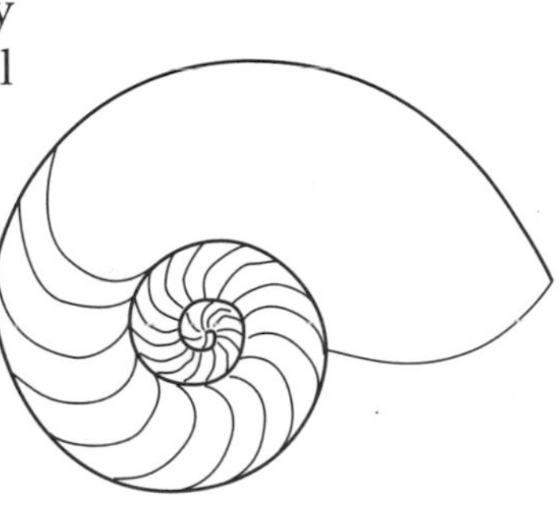

   Gnomonic growth can be modeled with many different shapes. Start with a square, a rectangle, or an isosceles right triangle, and construct a sequence of at least six figures to model the spiraling of gnomonic growth, as shown in these examples.

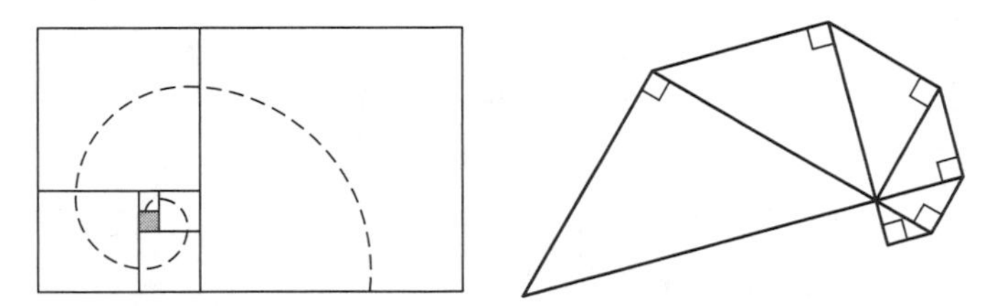

4. Each of the five tips of a pentagram is a **golden triangle,** an isosceles triangle in which the ratio of the length of a leg to the length of the base is the golden ratio. Use your construction from Step 4 to construct a golden triangle.

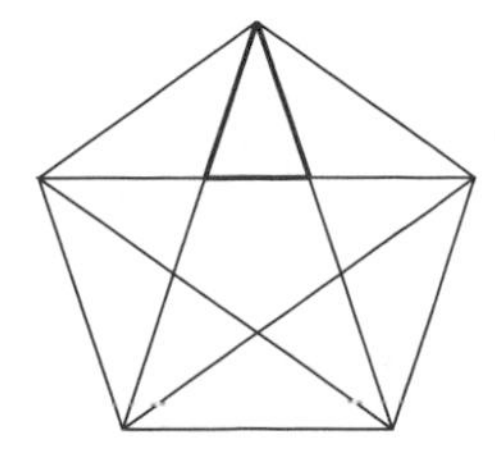

PROJECT • The Geometry of Baseball

contributed by Carolyn Sessions

Students use geometry to lay out a baseball diamond. This project can supplement Lesson 12.4.

OUTCOMES

1. The pitcher stands in front of the line from first to third base. By the Isosceles Right Triangle Conjecture, the distance from the line to home plate is $45\sqrt{2}$ ft, or about 63.64 ft, which is greater than the distance from the pitcher to home plate.

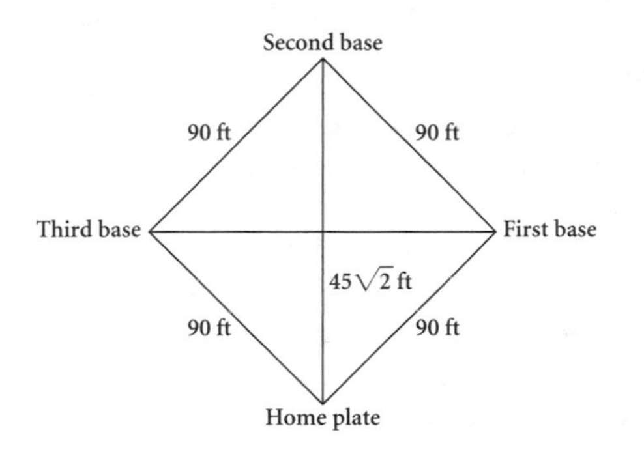

2. a. By the Isosceles Right Triangle Conjecture, the distance from home plate to second base is $90\sqrt{2}$ ft, or about 127.28 ft. This is approximately 127 ft $3\frac{2.8}{8}$ in., which is very close to the official measure. The official method of finding first base is equivalent to the SSS Congruence Conjecture, so it is correct.

   b. Sample answer: Although the official measure of the distance between home plate and second base is slightly off the actual length of the diagonal, it doesn't require the measure of any angles, so it is probably easier than measuring 90 ft around the sides of the square.

   c. The distances are specified the way they are because that is the division used on rulers measuring feet and inches.

   d. Sample method: Use a tape measure to find the distance from home plate to second base. Cut two 90 ft strings, attaching one to home plate and one to second base. Holding the ends of both, walk away from the line until both strings are taut and their ends are touching.

   e. The official distance between second base and home plate is 127.28125 ft. Using the Law of Cosines, the angle at first base, F, is

$$127.28125^2 = 90^2 + 90^2 - 2(90)(90)\cos F$$

$$F = \cos^{-1}\frac{127.28125^2 - 2(90)^2}{-2(90)^2} = 90.0018271°$$

So, the lines connecting first base to second base and first base to home plate are very nearly perpendicular, but not quite.

3. If the pitcher throws the ball at 90 mi/h, or $(90\text{ mi/h})\left(\frac{5280\text{ ft}}{1\text{ mi}}\right)\left(\frac{1\text{ h}}{60\text{ min}}\right)\left(\frac{1\text{ min}}{60\text{ s}}\right) = 132$ ft/s, it takes the ball $\frac{60.5\text{ ft}}{132\text{ ft/s}} \approx 0.458$ s to reach home plate. If the pitcher throws the ball at 60 mi/h, or 88 ft/s, it takes the ball about 0.688 s to reach home plate.

4. a. If a batter stands at the back of the batter's box, he or she has slightly longer to estimate the speed and height of the ball.

   b. For a 90 mi/h pitch, it takes the ball $\frac{57.5\text{ ft}}{132\text{ ft/s}} \approx$ 0.436 s to reach the front of the batter's box and $\frac{63.5\text{ ft}}{132\text{ ft/s}} \approx 0.481$ s to reach the back of the batter's box, a difference of about 0.045 s.

   For a 60 mi/h pitch, it takes the ball $\frac{63.5\text{ ft}}{88\text{ ft/s}} \approx$ 0.653 s to reach the front of the batter's box and $\frac{63.5\text{ ft}}{88\text{ ft/s}} \approx 0.722$ s to reach the back of the batter's box, a difference of about 0.069 s.

   c. Sample answer: The time delay gained by standing at the back of the batter's box is very small—less than a tenth of a second—but that might make the difference between hitting the ball and not hitting it. One disadvantage would be that you have slightly farther to run to first base.

TECHNOLOGY EXPLORATION • Polar Graph Designs

Students use graphing calculators to experiment with polar graphs. This project can supplement Chapter 12. You might use it with the Exploration Trigonometric Ratios and the Unit Circle. Students could also complete this activity using geometry software.

All screen shots are shown with range $0° \leq \theta \leq 360°$ and θ-step = 1. The graphing ranges given are all approximately square.

MATERIALS

- graphing calculators

GUIDING THE ACTIVITY

Step 1:

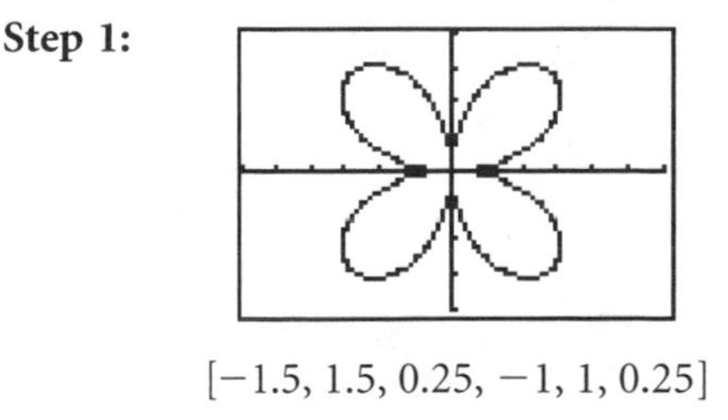

[−1.5, 1.5, 0.25, −1, 1, 0.25]

Step 2: n determines how many "petals" the daisy has. If n is even, the daisy has $2n$ petals. If n is odd, the daisy has n petals.

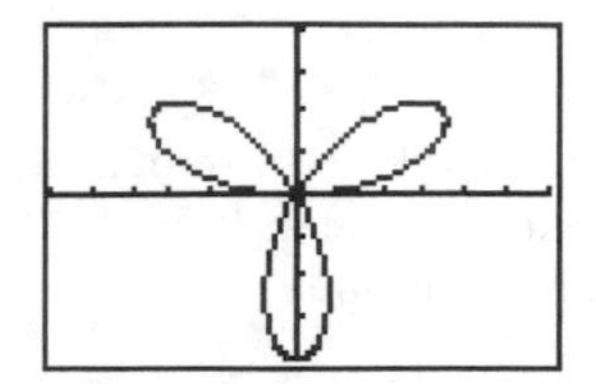

$[-1.5, 1.5, 0.25, -1, 1, 0.25]$

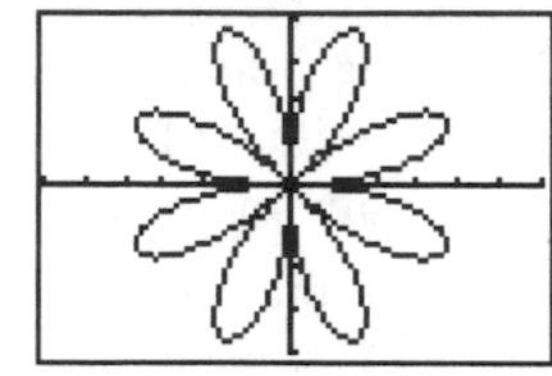

$[-1.5, 1.5, 0.25, -1, 1, 0.25]$

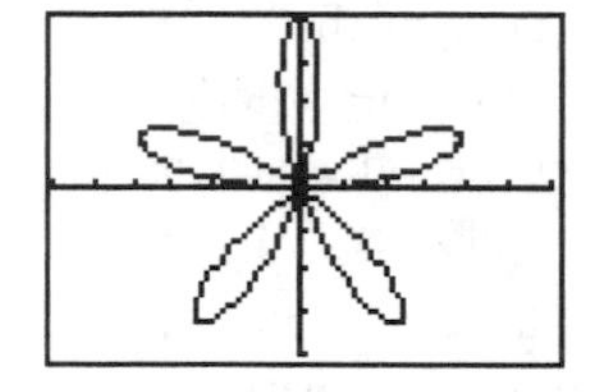

$[-1.5, 1.5, 0.25, -1, 1, 0.25]$

Step 3: a determines how long the petals are.

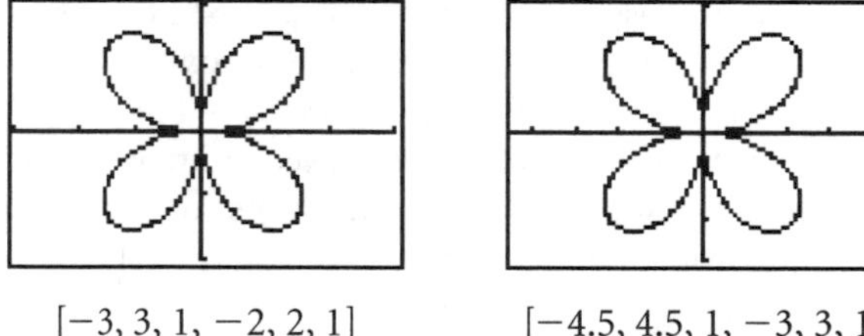

$[-3, 3, 1, -2, 2, 1]$

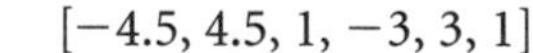

$[-4.5, 4.5, 1, -3, 3, 1]$

Step 4: The cosine graphs are the same as the sine graphs except that they are rotated 90° clockwise.

Step 5: Possible graphs:

$r = \theta$

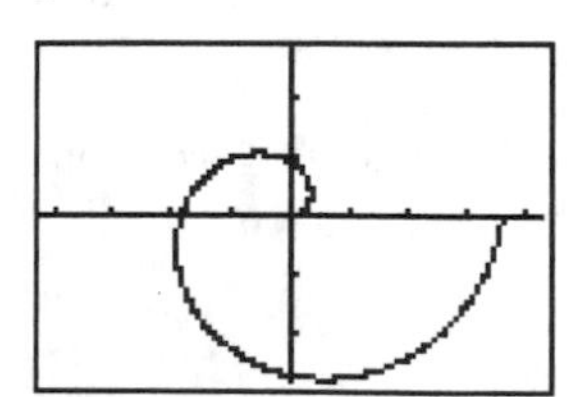

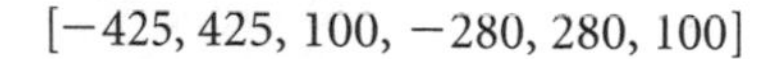

$[-425, 425, 100, -280, 280, 100]$

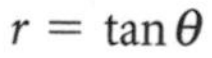

$r = \tan\theta$

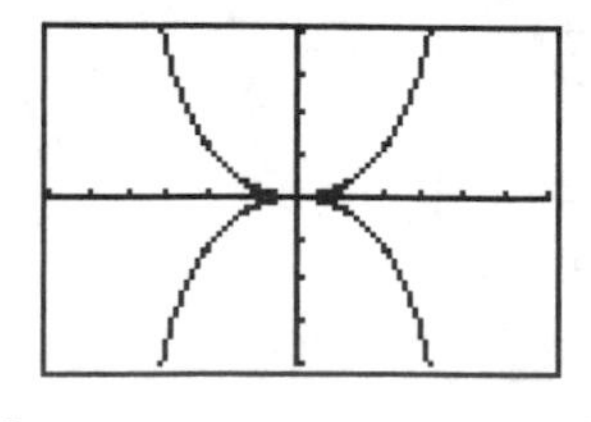

$[-1.5, 1.5, 0.25, -1, 1, 0.25]$

PROJECT • A Gothic Cathedral

Students review Chapters 0–12 by creating a "cathedral of geometry," using a stained-glass effect. If your classroom has windows, especially windows facing east or west, this project can be a fun way to screen out the light. If your classroom does not have windows, you might use the instructions to make a "quilt of geometry," with each section of glass corresponding to a quilt square.

Materials

- butcher paper
- large chalkboard compasses
- tracing paper
- markers
- Sample Design worksheet, *optional*

Instructions

1. Create a scale drawing of the windows in your classroom and construct the basic "lead" line design you want for each window. The Sample Design worksheet shows one possibility. You might have the designs within the circles in the top two rows feature types of geometric art found in Chapter 0: op art; line, circle, and knot designs; Islamic tile designs; and mandalas; and you could have the design within each circle of the lower two rows focus on one of the chapters from Chapters 1–12.
2. Divide your classes into as many groups as you have window sections, and assign one section to each group. Have each group choose a design that for them exemplifies the topic or chapter they're assigned. They should add their personal touches within the parameters you set so that the complete set of windows will have a unified look, theme, and color scheme.
3. For a quick and easy medium that produces a very colorful effect, have the students construct the first drafts of their designs with large chalkboard compasses on sheets of butcher paper; plan the details that personalize their artwork; and carefully transfer everything to large sheets of tracing paper. Then they can use markers to color the regions and finish the lead lines.
4. Attach the students' work to the windows with masking tape and enjoy the result!

PROJECT
The Geometry of Baseball

In this project you'll explore some of the many geometry concepts involved in baseball. Answer each question in detail, using geometry to justify your answers. Draw diagrams to illustrate your answers as necessary.

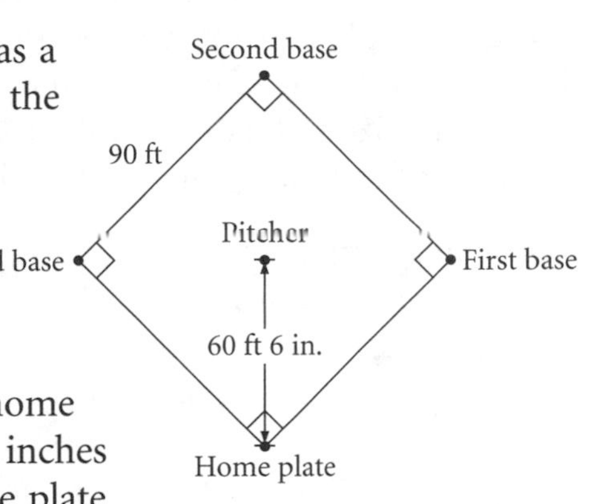

1. **Location of the Pitcher** A baseball diamond is officially defined as a square with side length 90 ft. The pitcher throws the ball to the batter from a position 60 ft 6 in. away from home plate. Is the pitcher's position in front of, on, or behind the line from first to third base? Draw a diagram and use mathematics to justify your answer.

2. **Laying Out the Bases** According to the 1992 official handbook instructions for laying out the field, "when the location of home plate is determined, use a steel tape to measure 127 feet, $3\frac{3}{8}$ inches in the desired direction to establish second base. From home plate measure 90 feet towards first base. From second base measure 90 feet towards first base. The intersection of these lines locates first base."

   a. Is the official method used to locate first base correct?

   b. Is the official method reasonable?

   c. Why do you think the instructions specify the distances as they do?

   d. Describe a technique that could be used to carry out the instructions.

   e. Are the lines connecting first base to second base and first base to home plate perpendicular?

3. **Pitching Speed** If the pitcher throws the ball at 90 mi/h, how long does it take the ball to reach home plate? How long does it take if the pitcher throws the ball at 60 mi/h?

4. **Where to Stand in the Batter's Box** The batter's box extends 3 ft behind and 3 ft in front of the center of home plate.

   a. Sometimes coaches advise batters to stand at the back of the batter's box when facing exceptionally fast pitchers. Why?

   b. For a 90 mi/h pitch, how much longer does it take the ball to get to a batter standing at the back of the batter's box than it does to get to a batter standing in the front of the box? How much longer for a 60 mi/h pitch?

   c. Discuss the advantage of standing as far as possible from the pitcher. Does this position have any disadvantages?

TECHNOLOGY EXPLORATION

Polar Graph Designs

You are used to graphing points on a Cartesian coordinate system, in which the location of a point is described by its distance from the x-axis and its distance from the y-axis. But this is not the only way to locate points on a plane. In **polar coordinates,** the location of a point is described by its distance from the origin, denoted r, and its angle with the positive horizontal axis, θ. Just as Cartesian graphs often show y as a function of x, polar graphs often show r as a function of θ.

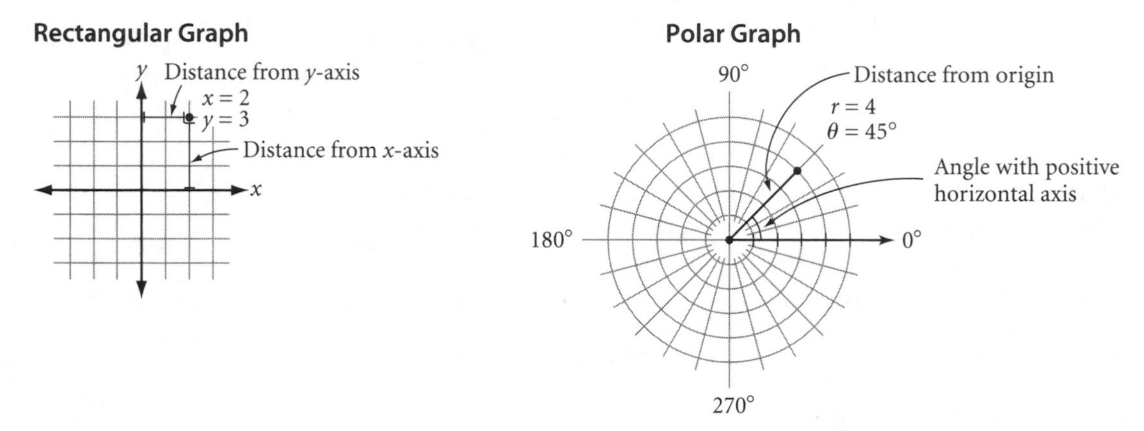

The trigonometric ratios you've been studying can be graphed as polar functions. These functions have graphs that are similar to daisy designs, though the curves are not arcs of circles. You'll study the mathematics of polar functions in a later course, but in the meantime, this activity can give you an idea of the role mathematics plays in design. You will need a graphing calculator for this activity.

Activity: Graphing Daisies

Step 1 Set your calculator to polar graphing mode. Graph the function $r = \sin 2\theta$. Use a range of $0° \leq \theta \leq 360°$. Sketch the graph in your notebook.

Step 2 Now graph and sketch $r = \sin 3\theta$, $r = \sin 4\theta$, and $r = \sin 5\theta$. In the general form of this function, $r = \sin n\theta$, what does the n tell you?

Step 3 Graph $r = 2\sin 2\theta$ and $r = 3\sin 2\theta$. In the general form of this function, $r = a\sin n\theta$, what does the a tell you?

Step 4 Replace sine with cosine in each equation in Steps 1–3, and graph the equations. How are these graphs similar to the sine graphs? How are they different?

Step 5 Experiment with other polar functions. Try functions with the tangent ratio and different types of functions you've learned about. What different shapes can you make?

PROJECT

A Gothic Cathedral

Sample Design

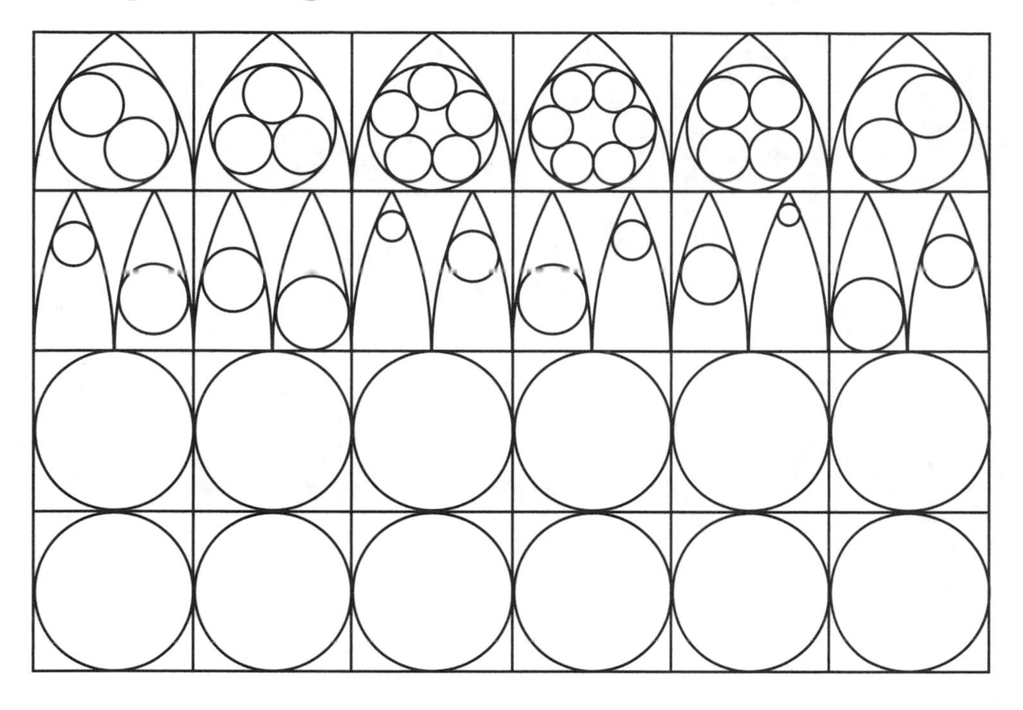

PROJECT • Proof by Mathematical Induction

contributed by Eberhard Scheiffele and Masha Albrecht

Students learn the process of mathematical induction. This is a very challenging topic. You might want to reserve this project for your mathematically ambitious students. Make sure students understand that mathematical induction is a method of proof and therefore a form of deductive reasoning—it's easy to get the *induction* in *mathematical induction* confused with the *inductive* in *inductive reasoning*. This project can supplement Lessons 13.1 and 13.2.

Outcomes

Example proof:

$$\frac{m(m+1)}{2} + (m+1) \stackrel{?}{=} \frac{(m+1)((m+1)+1)}{2}$$

$$\frac{m(m+1)}{2} + \frac{2(m+1)}{2} \stackrel{?}{=} \frac{(m+1)(m+2)}{2}$$

$$\frac{(m+1)(m+2)}{2} = \frac{(m+1)(m+2)}{2}$$

1. The sum of the first n odd numbers is n^2.

I. $1 = 1^2$

II. **Given:** $1 + 3 + 5 + \cdots + (2m - 1) = m^2$

Show: $1 + 3 + 5 + \cdots + (2m - 1) + (2m + 1) = (m + 1)^2$

Proof: Substitute and show both sides are equal.

$$m^2 + (2m + 1) \stackrel{?}{=} (m + 1)^2$$

$$m^2 + 2m + 1 = m^2 + 2m + 1$$

2. The sum of the first n even numbers is $n(n + 1)$.

I. $2 = 1(1 + 1)$

II. **Given:** $2 + 4 + 6 + \cdots + 2m = m(m + 1)$

Show: $2 + 4 + 6 + \cdots + 2m + 2(m + 1) = (m + 1)((m + 1) + 1)$

Proof: Substitute and show both sides are equal.

$$m(m + 1) + 2(m + 1) \stackrel{?}{=} (m + 1)((m + 1) + 1)$$

$$(m + 1)(m + 2) = (m + 1)(m + 2)$$

3. There is a total of $\frac{n(n-1)}{2}$ handshakes.

I. $0 = \frac{1(1-1)}{2}$

II. **Given:** $0 + 1 + 2 + \cdots + (m - 1) = \frac{m(m-1)}{2}$

Show: $0 + 1 + 2 + \cdots + (m - 1) + m = \frac{(m+1)((m+1)-1)}{2}$

Proof: Substitute and show both sides are equal.

$$\frac{m(m-1)}{2} + m \stackrel{?}{=} \frac{(m+1)((m+1)-1)}{2}$$

$$\frac{m^2 - m + 2m}{2} \stackrel{?}{=} \frac{m(m+1)}{2}$$

$$\frac{m^2 + m}{2} = \frac{m^2 + m}{2}$$

PROJECT • Writing a Logic Puzzle

contributed by Eberhard Scheiffele

Students learn to write logic puzzles. This project is fairly difficult and may be better for more advanced students. Students need to have completed the Explorations Sherlock Holmes and Forms of Valid Reasoning (Chapter 10) and Two More Forms of Valid Reasoning (Chapter 11).

Outcomes

- Student solves the project problem: Barbara is the girl, Isabelle is the dog, Marguerite is the cat, and Yuschka is the mouse.
- Student writes a complete logic puzzle with one logical solution.

Extra Credit

- Student proves his or her solution to the project problem is the only correct solution.

PROJECT

Proof by Mathematical Induction

Mathematical induction is a special procedure you can use to prove that a property holds true in an infinite sequence of cases—for example, for all positive integers. The procedure involves two steps. Here is one way you can think of the process.

Suppose you need to climb a ladder with infinitely many rungs. You want to be sure you eventually get to every rung. To be sure of this, you need to satisfy both of these conditions:

I. You know you can climb onto the first rung of the ladder.

II. Each time you are on a rung of the ladder, you know you can get to the next rung.

Can you see how these two conditions guarantee that you can climb the ladder forever? Here is another example.

Have you ever made a chain of dominoes? If you haven't ever made a domino chain, find someone who can describe one to you. Now imagine a domino chain that goes on forever and uses infinitely many dominoes. You plan to knock down this domino chain, but you want to be sure that you can knock over every single domino. One way to be sure is to satisfy both of these conditions:

I. You are sure you can knock over the first domino.

II. Each time a domino falls over, you are sure it will knock over the next domino.

In part I in each example, you need to show that a condition holds for the first possible case. In part II, you need to show that if the condition holds in any given case, it also must hold for the very next case after that.

For mathematical properties, proof by induction looks like this:

To prove that all positive integers n have some given property, prove these two conditions:

I. The number 1 has the property.

II. If a number m has the property, then the number $m + 1$ must also have the property.

Can you see how these two conditions are similar to the conditions in the examples of the domino chain and the ladder?

Now let's use mathematical induction to prove a conjecture about positive integers.

(continued)

Example

Look at Exercise 15 from the Chapter 2 Review on page 141 of your book. In this problem you found a pattern for adding sequences of positive integers. From this problem you can conclude that

$$1 + 2 + 3 + \cdots + n = \frac{n(n+1)}{2}$$

Although you probably already believe that this formula works for any positive integer, you haven't proved it yet. Use mathematical induction to prove this conjecture.

Solution

I. First we need to prove that the conjecture holds for $n = 1$. A sequence of length 1 that begins with 1 has only the number 1 in it. This means we need to prove that

$$1 \stackrel{?}{=} \frac{1(1+1)}{2}$$

Is this equation true?

II. Now we need to prove that if our conjecture works for some number m, then it also works for $m + 1$. Here is the setup for our proof. Notice that m is replaced with its successor, $m + 1$, in the show statement.

Given: $1 + 2 + 3 + \cdots + m = \frac{m(m+1)}{2}$

Show: $1 + 2 + 3 + \cdots + m + (m+1) = \frac{(m+1)((m+1)+1)}{2}$

Proof: The first m terms of the sequence on the left in the show statement are exactly the same terms as the sequence in the given statement, so we can substitute:

$$\frac{m(m+1)}{2} + (m+1) \stackrel{?}{=} \frac{(m+1)((m+1)+1)}{2}$$

Brush up on your algebra skills! Complete the proof by showing that both sides of the equation are equal.

Now you'll write your own proofs using mathematical induction. For each problem, copy and complete the statement, then prove the statement using mathematical induction. Set up both steps of each proof carefully.

1. The sum of the first n odd numbers is ________________.
2. The sum of the first n even numbers is ________________.
3. Suppose there are n people at a party. Also suppose every person shakes hands with every other person. There is a total of ________________ handshakes. (See the Investigation Party Handshakes on pages 108–110 of your book.)

PROJECT

Writing a Logic Puzzle

In the Explorations Sherlock Holmes and Forms of Valid Reasoning in Chapter 10 and Two More Forms of Valid Reasoning in Chapter 11, you've studied the basic principles of logic. In this project you'll apply these principles to write puzzles whose solutions depend mainly on a correct application of logic or deductive reasoning. Consider the following example:

The theft was committed by one of Rose, Charles, or Steve. If Charles had committed the theft, Holmes would know about it. Neither Holmes nor Watson knows who committed the theft. If Steve did not commit the theft, Rose is innocent.

The information given is sufficient to tell you who committed the theft. You only need to apply logic. One way to think about it is to realize that to assume Charles or Rose committed the theft leads to a contradiction. If Charles had committed the theft, Holmes would know about it. But Holmes doesn't know who committed the theft. This is a contradiction, so it is impossible that Charles is the criminal. If Rose had committed the theft, then by *Modus Tollens* you would know it is not true that Steve did not commit the theft. This would mean that both Steve and Rose are the criminals. This is a contradiction to the first statement that one of the three—Rose, Charles, or Steve—committed the theft. The last remaining choice is that the thief is Steve. Now you should, of course, make sure that assuming Steve committed the theft does not lead to a contradiction.

If you let T represent "is the thief" and let K represent "knows who committed the theft" (and let R represent Rose, and so on), you can write the given information in the example as

$T(R)$ or $T(C)$ or $T(S)$

$T(C) \rightarrow K(H)$

$\sim K(H)$ and $\sim K(W)$

$\sim T(S) \rightarrow \sim T(R)$

Logic alone allows you to conclude that $T(S)$ is true and hence Steve is the thief.

Now try this puzzle:

During lunch, the students are discussing the adventures of Barbara, Isabelle, Marguerite, and Yuschka. The teacher, who listened for a while, asks them, "Who are you talking about?" The oldest student answers: "We are talking about a girl, a cat, a dog, and a mouse." The teacher replies, "But who is what?" The students want to challenge the teacher, so they answer, "If Yuschka is not the mouse and Isabelle is not the girl, Marguerite is the dog. If Barbara is not the cat, then, if Yuschka is not

(continued)

the girl, Isabelle is the dog. At least one of the three following statements is correct: Marguerite is the cat, Barbara is the mouse, Yuschka is the dog. If neither Marguerite nor Barbara is the girl, Isabelle is the dog."

"Stop! That's enough for me to know who is what," the teacher interrupts.

Who is the girl, the cat, the dog, and the mouse? (*Challenge for extra credit:* Prove that your answer is the only possible answer.)

Here are some steps you might follow to make up similar puzzles:

Make up a story in which you use names of persons, animals, or things. It's best to start with a specific assignment of the names to the objects in mind. Let's say, for example, you want to make up a story about your hamster Rudi, your snake Till, and your turtle Joel. Now you have to make up sentences with logical connectives (*and, or, if-then, not,* and so on) that are true under your assignment of names to objects. Obviously it would be too easy to just use "Rudi is the hamster, Till is the snake, and Joel is the turtle." So you might try (1) "If Joel is not the hamster, then Till is the snake."

Always check to make sure your sentence is true. You need to create sentences that are false under some of the other possibilities (of assignments). For example, the sentence (2) "Joel is not the hamster or Till is the snake" excludes all assignments in which Joel is the hamster and Till is not the snake.

When you have enough sentences to exclude all the other possibilities, you are finished. For example, if you add (3) "Joel is the turtle or Rudi is the snake," then sentences (1)–(3) imply that Joel is the turtle. Why? Check that assuming Rudi is the snake leads to a contradiction. But then (1) implies that Till is the snake. Then, of course, Rudi has to be the hamster. So sentences (1)–(3) allow only one solution.

Now make up your own puzzle and exchange it with another student. Don't forget to keep a note with the correct solution. It's easiest to start with only two different names and then make up more difficult puzzles later. You may even want to put in some additional statements that are not pertinent to the solution to make the puzzle more difficult to solve.

Comment Form

Please take a moment to provide us with feedback about this book. We are eager to read any comments or suggestions you may have. Once you've filled out this form, simply fold it along the dotted lines and drop it in the mail. We'll pay the postage. Thank you!

Your Name ______________________________

School ______________________________

School Address ______________________________

City/State/Zip ______________________________

Phone ____________________ Email ____________________

Book Title ______________________________

Please list any comments you have about this book.

____

____

____

____

____

____

____

____

____

____

Do you have any suggestions for improving the student or teacher material?

____

____

____

____

____

____

____

To request a catalog or place an order, call us toll free at 800-995-MATH or send a fax to 800-541-2242. For more information, visit Key's website at www.keypress.com.

Fold carefully along this line.

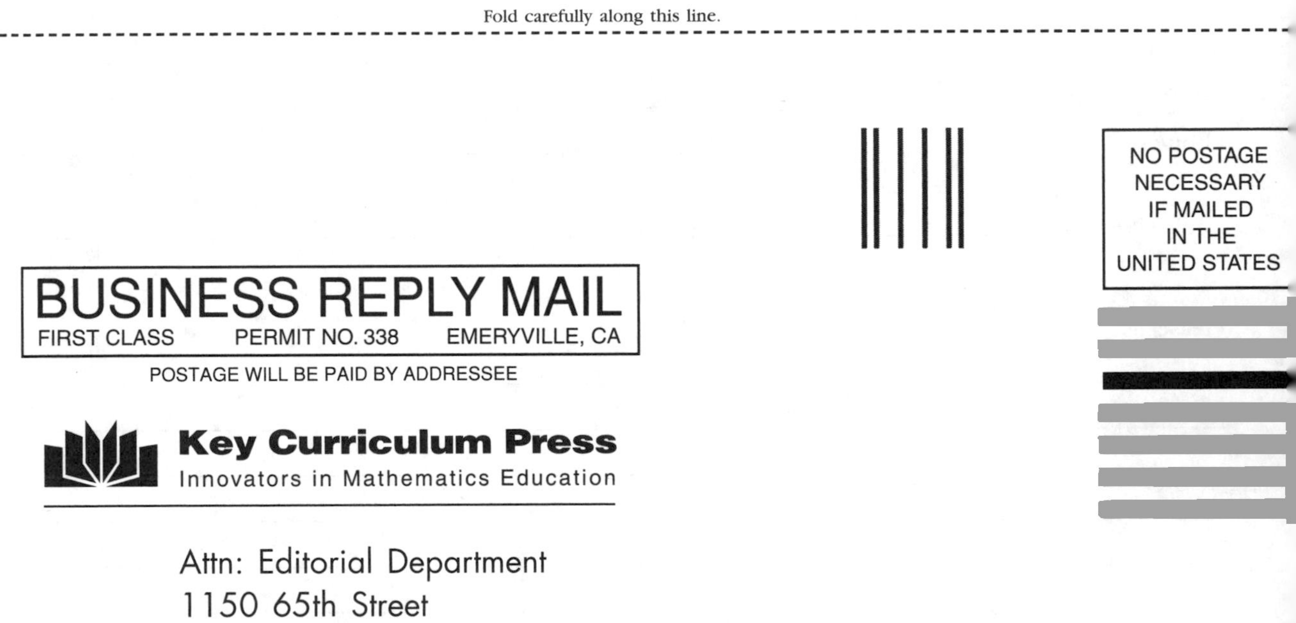

Fold carefully along this line.

Comment Form

Please take a moment to provide us with feedback about this book. We are eager to read any comments or suggestions you may have. Once you've filled out this form, simply fold it along the dotted lines and drop it in the mail. We'll pay the postage. Thank you!

Your Name ____________________

School ____________________

School Address ____________________

City/State/Zip ____________________

Phone ____________________ Email ____________________

Book Title ____________________

Please list any comments you have about this book.

____

Do you have any suggestions for improving the student or teacher material?

____

To request a catalog or place an order, call us toll free at 800-995-MATH or send a fax to 800-541-2242. For more information, visit Key's website at www.keypress.com.

Fold carefully along this line.

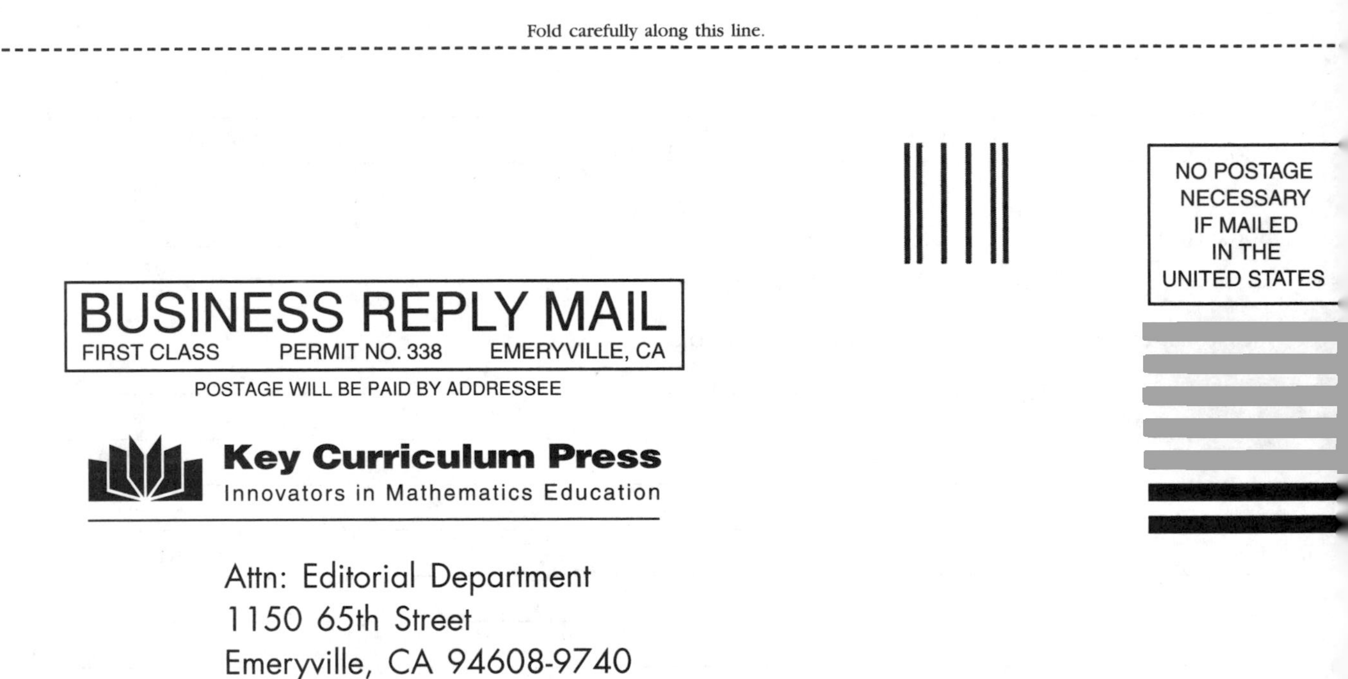

Fold carefully along this line.